AUSTRALIA IN TRUST

Published jointly by William Collins Pty Ltd and the
Australian Council of National Trusts.

William Collins Pty Ltd,
55 Clarence Street,
Sydney, NSW 2000

Australian Council of National Trusts
PO Box 1002
Civic Square, ACT 2608

Copyright 1985 Australian Council of
National Trusts

All rights reserved. No part of this publication may
be reproduced or transmitted in any form or by any
means, electronic or mechanical including photocopying,
recording or by any information storage and retrieval
system without written permission from the publisher.

National Library of Australia
Cataloguing-in-Publication Data:

Australia in trust.

ISBN 0 00 217541 X.

1. Historic sites – Australia – Conservation and
restoration – Addresses, essays, lectures.
2. Historic buildings – Australia – Conservation
and restoration – Addresses, essays, lectures.
3. Environmental protection – Australia – Addresses,
essays, lectures. I. Australian Council of National
Trusts. II. Title: Heritage Australia.

363.6'9'0994

Designed by Ross Buxton
Printed and bound at Griffin Press Limited, Adelaide

AUSTRALIA IN TRUST
A Selection of the Best Writings from 'Heritage Australia'

AUSTRALIAN COUNCIL OF NATIONAL TRUSTS
IN ASSOCIATION WITH WILLIAM COLLINS PTY LTD

Contents

Foreword   *H.E. E. G. Whitlam, Australian Ambassador to UNESCO*   6

Introduction   *Rodney Davidson, OBE, Chairman of the Australian Council of National Trusts*   7

Towards the bicentennial landscape   *Judith Wright*   8

The World Heritage List   *Ralph Slatyer*   18

Lord Howe Island   *Chris Murray*   24

Kakadu — man and landscape   *Allan Fox*   34

Movables of heritage significance   *Rosemary Hassall*   51

Andrew Lenehan — Sydney cabinetmaker   *Kevin Fahy*   57

South West Tasmania   *Josephine Flood*   62

Australian themes in stained glass   *Beverley Sherry and Douglass Baglin*   68

War memorials in our landscape   *K. S. Inglis*   75

Cemeteries — their value and conservation   *James Semple Kerr*   80

Rouse Hill House   *James Broadbent and Douglass Baglin*   90

Sydney's first skyscrapers   *Len Fox*   98

Twentieth century buildings   *Peter Spearritt*   104

Mulwala homestead complex   *Peter Freeman*   110

Harrisford, Parramatta   *Clive Lucas*   115

Surf and steel   *Rosemary Auchmuty and Peter Spearritt*   120

Lyndhurst — a battle won   *Clive Lucas*   130

Edmund Blacket's church architecture   *Joan Kerr*   135

Vaucluse House   *Peter Watts*   146

High country huts *Klaus Hueneke*	150
Garden Island *Eric J. Martin*	156
Carpenter's decoration *Kate Blackmore*	162
First Government House, Sydney *Helen Proudfoot*	167
Port Arthur Historic Site *Brian J. Egloff*	173
Rippon Lea *Miles Lewis*	179
The restoration of Clarendon *G. M. W. Clemons*	182
Along the Ghan track *John Wood*	188
The art of the carpenter *Frank Bolt*	198
Adelaide's stone buildings *Alan Spry*	204
The Queensland house *Richard Allom*	212
Palma Rosa, Brisbane *Peter Marquis-Kyle*	220
The Palace Hotel, Perth *Penny Grose and R. McK. Campbell*	224
Yarloop, Western Australia *Penny Grose*	228
Historic bridges of Australia *Colin O'Connor*	234
The rock paintings of Cape York *Josephine Flood*	238
South East Queensland coal mines *Raymond Whitmore*	246
Archaeology of the Burrup Peninsula *Nicholas Green*	252
The Register of Significant Trees *Eve Almond and Peter Lumley*	256
Historic public gardens, Perth *Oline Richards*	261
Notes and references	268

Foreword

The range and quality of the articles republished in this book explain the interest which Australian Governments and international organizations are taking in Australia's heritage, cultural and natural.

Without the members and supporters of our National Trusts it would not have been possible for my Government to assemble the public-spirited members of the Hope Committee on the National Estate and the Pigott Committee on Museums and National Collections. Without such support it would not have been possible to secure public support for Federal initiatives to inscribe the Great Barrier Reef Marine Park, Kakadu National Park, Willandra Lakes Region, Lord Howe Island and, finally, Western Tasmania Wilderness National Parks on the World Heritage List. Nor would the States, with New South Wales in the vanguard, have yet preserved other natural and Aboriginal sites and restored the cultural properties built by the first British settlers. The Australian Museum, Sydney, has completed the Unesco-funded inventory of Polynesian and Micronesian artefacts in Australia; it is the most thorough inventory of cultural property achieved for any region in the world.

Public support has led to Australia's ratification of the three Unesco Conventions for the Protection of Cultural Property in the Event of Armed Conflict, on Prohibiting and Preventing the Illicit Import, Export and Transfer of Ownership of Cultural Property and concerning the Protection of the World Cultural and Natural Heritage. This in turn has led to Australians now being members of the governing bureaux of the World Heritage Committee and of the three bodies which advise the Committee, the International Centre for the Preservation and Restoration of Cultural Property (Rome Centre), the International Council of Monuments and Sites and the International Union for Conservation of Nature and Natural Resources. This year it is likely that Australia will be elected to Unesco's Intergovernmental Committee for Promoting the Return of Cultural Property to its Countries of Origin or its Restitution in Case of Illicit Appropriation.

Readers of this book in Australia and overseas will be reminded and re-assured of how much Australia's heritage and reputation have depended and will depend on the efforts of the diligent and dedicated men and women who support Australia's National Trusts.

Gough Whitlam
Australian Ambassador to UNESCO

Introduction

In 1988 Australia celebrates the bicentenary of its European settlement. That settlement has added a European and Asian heritage to the nation's Aboriginal and natural heritage. The total is what creates the Australian environment today. Representative parts are being held 'in trust' for the benefit of current and future generations: in the National Trust's view many more components need protection. To be good custodians we must be sure that as Australia develops, progress is carefully linked with the preservation of significant aspects of our past. That past is an important influence on, and reference point for, the present and the future. It is a source of pride for Australians and a major source of interest for visitors.

A starting point for understanding Australia's heritage is to list and evaluate the components of our nation's heritage. The National Trust movement has been undertaking that work for nearly forty years. More recently, Government has become involved. The Australian Heritage Commission, and those States which have statutory bodies concerned with preservation, are now also compiling registers. It is a vast task being undertaken with limited human and financial resources, but the work is well under way. It is also an ongoing task, and as the surveying continues, previously unrecognised components of our National Estate are discovered and new parts of our known heritage are listed. At last, good researched inventories exist and continue to be developed. In this work, Governments continue to draw heavily on the expertise and work of the National Trust.

In addition to this development of registers, there is a need for continual vigilance. Although a great deal has been lost through destruction or neglect, fortunately a great deal has been saved.

The Australian Council of National Trusts is the Federal coordinating body for what is now Australia's largest voluntary conservation organisation. Forty years ago a small group of citizens met in a private home in Sydney, and two years later the National Trust in Australia was launched in New South Wales. Twenty years later the Council was formed to serve the national interests of the Australian Trusts now established in each of the country's eight States and Territories. Today, the Trusts have a national membership of 80,000 people, own around 300 properties and through their own energies spend nearly as much annually on the preservation of Australia's National Estate as does the Australian Government through its National Estate Grants Program.

I hope that as you read this book you will appreciate the heritage we have and the need to take carefully planned action to conserve and protect its important elements. They have been handed down to us and we, as a nation, should preserve and conserve them either by public action or by measures designed to ensure their continuity in private ownership.

The Trust wishes to thank all those whose work, knowledge and expertise have made the book possible.

Rodney Davidson
Chairman – Australian Council of National Trusts

Towards the bicentennial landscape

Judith Wright
© *Judith Wright, 1982*

Judith Wright McKinney D.Litt. (Hon.), F.A.H.A., is one of Australia's foremost poets, authors and conservationists. Her name is synonymous with the nation's endeavour to preserve the environment.

Prophecy is notoriously a dangerous business, and those who accept the challenge of indulging in it in their own countries cannot say they have not been proverbially warned. But, in the first years of the final decade of our second century of occupation of Australia, it may not be so difficult to foresee the main aspects of the landscape in which that bicentennial will be celebrated. The events of the past, and the trends in land-use we have already established, are scarcely likely to be changed or reversed in the remaining six years of the century.

However, there is no secure baseline from which we can reckon. Records of what the country looked like, what is distribution of forests, plants, animals and insects were on our intrusion, are at best scrappy, and in many places non-existent. Though such observers as Banks, Brown and Dampier examined the flora of the coastal areas they saw, the main reports on inland flora and fauna were made by passing land-explorers, who naturally saw only what lay in their paths, and spent little time in observation. Even Leichhardt, whose interests were much wider than those of other explorers such as Mitchell and who was intrigued by the relationships rather than the mere taxonomy of species, could report only on those aspects he saw at the particular time he passed through.

Those who followed in the land-explorers' tracks were seldom qualified or interested in identifying the components of the plant and animal products of the country — it was sheep-pasture that concerned them, and even there they were inexpert.

Therefore, it was only as obvious and damaging changes in the land affected their enterprises that the Colonies enlisted scientific help. Such men as Frederick Bailey in Queensland saw only a landscape already drastically altered by decades of increasingly heavy grazing by sheep, cattle and horses; and, as botanists, they were unqualified to describe the aspects of the country which Leichhardt, that ecologist before his time, attempted to discover in his too-brief Australian years.

By the first centenary of white occupation, such changes were disquietingly visible. The imposition of European methods of land-use had altered the coastal plain on which the first settlements were established, from the beginning. Apart from the cutting down of forests, and the quarrying of stone, which began as soon as Sydney Town was set up, there was the landing of the first domesticated European animals, whose grazing habits and impact on the country were much more damaging than those of the only large indigenous mammals, the kangaroos.

Small farms were established on watercourse banks and alluvial flats which had previously supported only Aboriginal camps — camps which moved from place to place in accordance with the natural harvest of plant and animal food, used little fuel for their fires or materials for their shelters, weapons and implements, but managed their landscape so as to keep open the pasture areas on which their marsupial prey grazed. How

Towards the bicentennial landscape

Aerial view of saltbush country north of Broken Hill. (Photo: Colin Totterdell)

much impact their occasional firing of the grasses had, and how much of the landscape the first white arrivals saw was their creation, we cannot know. But fire is still a feature of the sclerophyll forests, and a regenerator of certain species while it controls undergrowth.

Elizabeth Macarthur's observation that 'the greater part of the country is like an English park' — that is, a nobleman's demesne kept clear of undergrowth, where deer might be easily hunted and find good pasture — is, as Bernard Smith points out, corroborated by the paintings of the Port Jackson Painter, who shows a 'countryside composed of gentle lawn-like foregrounds which slope down to a lake-like harbour backed by evenly rounded hills which are dotted with isolated clumps of trees and fringed with long belts of timber.'[1] This open landscape, without undershrubbery, seems almost certainly to have been as much the result of human action as the nobleman's deer-parks themselves.

The small farms of the settlement and of the near hinterlands of Windsor and Parramatta were a wholly new intrusion of such landscapes; those squared and dug enclosures in which the first wheat, maize and vegetables were planted and in which the virgin soils were broken for the first time exposed them to erosion in rain, drying in heat, and changes in the (still unknown today) original soil fauna and flora. The European imposition of strict boundaries and straight lines produced the first adumbration of a network of such imaginary, or fenced, lines which was, by the end of the century, to cover virtually all the desirable areas of the island continent. Such surveyed boundaries took as little account as possible of the natural contours, slopes, soils, aspects and ground cover of the Aboriginal landscape, and of its swamplands, watercourses and forests. Compared with the land-use patterns of Aboriginal days, which took ecological rather than mathematical laws as their base, such a network was ecologically farcical. It might cut through the centre of forests and wetlands, divide one bank of a river from the other and one section of a watercourse from the rest, and impose land-uses quite unsuitable to the terrain. It has resulted in a new kind of mosaic than the old mosaics of

Towards the bicentennial landscape

Top 'The goldmining areas of the 19th century were largely denuded of young trees'. Pictorial survey document (c. 1865) of the Energetic Gold Mining Company's works on Victoria Reef, Long Gully, Bendigo District, Victoria. (Print by courtesy of Mr and Mrs P. Dunn of Canberra)

Above Horse teams ploughing a wheat field in the 19th century. (Photo: Australian Information Service)

burned and unburned pasture, one in which cleared, grazed or cultivated cropland is marked out within surveyed lines which take little account of the capabilities of the land.

The first roads and tracks of the various Colonies set up in the following years were as a rule less puritanically laid down, for men on foot or on horseback, bullock-drays and carts, tended to take not the straight line between two points, but a more rational route dictated by the steepness of slopes, the location of fords and mountains, and the distance between reliable waterholes. Many country roads still approximate those early routes, though the surveyors have had their will — however expensively — in cities and towns and have spread the dictates of Roman roadbuilding wherever possible.

When the coastal shelf on which the penal settlements were built began to prove insufficient to supply the town with food, the Bathurst plains and the Hunter Valley were opened up, and flocks and herds began to cross the Dividing Range and spread out beyond it in defiance of the Colonial Office's instructions. While, within the Nineteen Counties, land grantees were constrained within the boundaries of their properties, it was a different story on the Liverpool Plains and in Australia Felix, where the writ of the surveyors did not run, and here it was the possession of water-frontages and pasture which set off early quarrels among the 'squatters' and between the shepherds who were occupying the country on their behalf. As the possession of access to water and water-frontages was a crucial factor for Aboriginal survival, the battles that followed were deadly.

The size of the early 'stations' was immense, and their boundaries tended to be laid down on something the same lines as Aboriginal land-use may have followed — for no squatter wanted the rough and mountainous ranges where little grass grew, and everyone wanted to centre his occupation on the best waterholes. Land Commissioners and surveyors later set such natural boundaries in more Roman order, so that when fencing of boundaries became necessary later, fencing teams cursed their way over rocky slopes and ranges and squatters cursed the resulting expense. Boundaries between colonies were equally authoritative in their straightness where this was possible, and once again the ecological facts of the land and its topography were ignored.

By this time, however, the flocks for whose benefit the country had been invaded had greatly changed the aspect of the 'outside country'. (Cattle did not increase as fast as sheep, and could find pasture where sheep could not; and their capacity to walk longer distances to water and to fend off dingo attacks allowed them to occupy the rough country of the ranges. Since beef was not exportable until meatworks were set up, many cattle occupied the scrubs and ranges and ran wild there.) The shepherded flocks required about one man to each 500 or so sheep, plus a hutkeeper; huts and yards occupied the banks of watercourses, generally on permanent waterholes, but could be and were moved from place to place in drought or when pastures were eaten out. Aboriginal attacks were a factor in early times, and shepherds seldom dared to take the flock far out, when alone, so that pasture composition and density along the waterfrontages suffered first. Trees were felled for huts, yards and cooking-fires, and the banks of streams suffered erosion and collapse not only from overgrazing but denudation of tree-cover. Such early changes during the first years of white management must have been considerable factors in the incidence of floods and the damage they caused.

This was increased by the draining of marshland, swamps and lagoons which seem to have bordered most watercourses in earlier times, taking the overflow of heavy rainfall, and releasing it slowly back as the main stream sank. Though little documented, this change was evidently noticed by the observant, such as Charles and Henry Barnard in Central

Queensland.² Henry, in an article written in the 1920s, noted that the fringing swamps and lagoons of the Lower Dawson area had largely vanished, drained by long cattle-pads of trampled soil; sheep-pads, which terrace slopes and lead to erosion, have been observed to have the same effect on shallow lagoons and marshland, in which both sheep and cattle search for food in drought. Charlotte May Wright, in her memoirs, observed also that the Hunter Valley's swamps and lagoons, which she remembered from her childhood in the 1850s and 60s, had disappeared by the 1920s when she wrote, and 'now there seem to be no lagoons and very few ducks.'³ Other observations on the plains of the northwest seem to indicate that the same pattern of change has occurred there, and no doubt it was much more general than has been realised.

These changes to watercourses and storages were paralleled by changes to soil structure. The complaints of waggoners and draymen in early days, taking their loads over country never compacted by hooves or wheels, were great. William Wiseman, Crown Land Commissioner for the Leichhardt District, explained in an offical dispatch that he would need a bullock rather than a horse-team for supplies, 'as the ordinary dray tracks require a long traffic on them before they are sufficiently hardened to allow the passage of a horse-team'. Deep spongy soils full of the humus of centuries were 'sufficiently hardened' by the hooves of sheep and cattle over the following century to bear wheels almost anywhere, and the bogs which also beset draymen and horsemen were no longer a problem.

Apart from the spreading of pastoral industry, the second major object of the users of land was the clearing of forest. Along the eastern coasts and wherever tall forest occurred, timber-getting was an early and lucrative industry. The red cedars which had been mercilessly logged in the east were virtually cut out by the 1860s in New South Wales; by 1888 they were growing scarce even in the extensive and heavy forests of Queensland; such timbers as kauri pine and the cabinet-woods of the rainforests soon followed. Virtually no control and no forestry practices were imposed on such early exploitation. The illusion — still to be found — that the heavier the forest and the taller the trees, the richer must be the soils under them, caused a great deal of valuable timber to be cleared and burned for farmland which within the next few decades often became derelict and weed-infested as the first fertility of ash and humus was exhausted; such forests often grew on soils quite unsuitable for farming and on slopes whose gradient was an invitation to erosion. Much of this land remains abandoned today (as may easily be seen by those who drive inland from the Gold Coast to view the ravines and hills round Beechmont and Springbrook).

In New South Wales and Victoria, the goldmining areas of the nineteenth century were largely denuded of tall young trees suitable for pit-props, and timber-getters for other purposes wasted immense quantities of valuable timber in times when 'trees the growth of centuries [were] plentiful and cost nothing.'⁴ The goldfields themselves were ravaged and left without any rehabilitation measures as miners rushed to the next bonanza; most remain in that state today, and the cyanide used in mining processes was destructive to wildlife and plantlife for many years after.

It was, however, this triumphant exploitation of the land's resources which provided the wealth men pointed to in complacent speeches during the celebration of the first centenary. At that time, in spite of the influx of miners from the early 'fifties onward, the Australian colonies still supported only about three million people, and there was great concern that this low population (even then centred on the coastal cities) would invite the descent of some Yellow Peril. The treaty between China and Britain was deplored, for it made it impossible to forbid Chinese immigration to the goldfields and elsewhere, and this was increasing in spite of heavy poll-taxes and a punitive customs duty on opium. Only an increase

Eucalypts cleared for pine in south east New South Wales. (Photo: Colin Totterdell)

in population, politicians warned, could prevent an impending invasion; and free immigration from Britain was the best remedy. For years, until the Land Acts divided the great pastoral runs, people had pointed out that those vast areas empty of almost everything but grass, sheep and cattle were an open invitation, and an industrious yeomanry to occupy the inland would not only discourage such an invasion but add to colonial wealth through land rents and sales.

But the trend which was already swelling the cities with refugees from the problems of the land seemed irresistible. Though the eastern Colonies had (with the exception of well-populated Victoria) spent much on assisted immigration, few of the British arrivals could put up with the hardships of the inland.

Two factors were optimistically quoted — the railways, which were spreading far into previously isolated country, would surely be the forerunners of large and flourishing centres of population to follow; and the new availability of small grazing farms on the cutting-up of those big runs. Sceptics who pointed to the inland's now notorious lack of surface water and recurring droughts were answered by optimists who declared that the new discovery of artesian water would solve the problem and, as the poet O'Dowd declared, provide 'a perfect zone of broadening green' to bless the 'utmost wilderness'. (*The Dominion of Australia: a Forecast.*) The bore put down at Blackall in Queensland was already attracting eager intending settlers on those empty grazing farms; and politicians, and even some scientists, were predicting that similar underground supplies existed everywhere in the island continent if one could only drill deep enough. And these new American drilling machines seemed capable of reaching any depth. The inland rivers, with their perplexing habit of disappearing without obvious outlet, must surely have been storing their waters for aeons in some providential underground reservoir now to be tapped by the triumphs of science and technology.

There were other problems, however, which the Centennial speech-makers generally ignored. The great pastures which had supported so many introduced flocks and herds were already failing. Both introduced weeds and native plants inedible to sheep and cattle were increasing; burrs and thistles had spread far along the stock-routes, carried by flocks and mobs, and scrubland of wattles and brigalow was spreading to encroach on the paddocks. The 1884 Land Act of Queensland, which had offered such scrublands at a peppercorn rental to those willing to clear them, was attracting few men to the task. Blackberries and sweetbriar were proving a problem on coastal and tableland watercourses and pastures. Worst of all, rabbits, once such welcome introductions, had already advanced up the Darling basin, eating grasses even down to the roots, killing trees and shrubs by nibbling off their bark, and were already reported to be inside the tardily built rabbit-fence intended to keep them out of western Queensland. Along the Moonie and Condamine, alarmed settlers were complaining of the difficulty of clearing flourishing growths of prickly-pear, and men who took up grazing farms were abandoning the job of chopping and burning the formidable plants.[5]

For Australia's native animals, as for her other original inhabitants, the years had brought disaster. The original distribution of plants and animals dependent on those plants is little known but the changes already visible must have affected many species. It is known that the bilbies early settlers had relished as 'native rabbit' were already vanishing from some of their haunts as introduced rabbits took over their burrows and food-supplies. The cessation of Aboriginal hunting had brought many changes; first, dingoes — whose litters had been a food source for Aborigines — increased greatly as they developed a taste for stray lambs and weakling sheep in the wake of the moving flocks, then a massive strychnine-baiting campaign against dingoes took also many other predators with it. When the dingoes were reduced, marsupials increased accordingly and their concentration on the pastures which were also the chief support of sheep frightened flock-owners and caused immense expense in ammunition and bonus payments for scalps under the Marsupial Acts.

Other effects were less noticeable to pastoralists, but must have occurred nevertheless as forests were felled and water-supplies altered in their nature. Bustards, always desirable additions to the spartan diet of the inland, were vanishing in many places; but under the cover of those advancing walls of prickly-pear scrub-turkeys flourished and death-adders and other snakes, now invisible to hawks and eagles, bred in plenty. The birds which had learned to eat the thorn-studded fruit — emus, currawongs and others — did well, and spread the seeds further — or so said the men who were battling against the pear. Other changes found few recorders.

The influence of droughts, floods, and inland hardship were made stronger by the erratic boom-and-bust swings in the pastoral economy, always hung on a long umbilical cord from European markets and demands. Soon the long depression of the 'nineties would drive from the land many of those who had first arrived and had so far held their land; then the great drought of the end of the century and the early years of the 1900s would add its influence. Absentee ownership, whether by banks and finance companies or by big and often overseas-owned pastoral firms, increased (and it was not often that such ownership was enlightened in its attitude to the needs of the land or knowledgeable in its problems). All this further increased the drain to the cities, where goods and services, education and markets offered far more attractive conditions than did the little country towns and townships. There would be no prosperous yeomanry to occupy the great stretches of the inland, and even those artesian waters often proved too salty and chemical-laden to support agriculture.

The second century of occupation changed this picture as tractors and other machinery were put to use in clearing and ploughing vast acreages for wheat. The inland once more altered past recognition as cropland was pushed far into the semi-arid zone. Unwise clearing of the land, as in the mallee country, resulted in encroaching dunes of sand and in the 'forties drought dust-storms sent red Australian topsoil far across the Pacific, but a second great industry was established in formerly pastoral country. From aircraft passing over the western plains, passengers could now contemplate the great network of straight lines into which the country had been surveyed, enclosing areas of red, pink or yellow soil or varying shades of cropland green. The vast new powers now at work on the land could clear hitherto resistant forests in less time than men had taken to cut scrub from their back paddocks, and the 'brigalow belt' of central and western Queensland vanished to be replaced by cattle-country and fields of sorghum and other crops. Such apparent triumphs restored many men's faith in the power of technology to tame a recalcitrant country.

It was possible now to build great dams and reservoirs with little labour. Such projects as the Ord Dam and the hydro-electric schemes in the Tasmanian river-wilderness country were much admired (though cost-benefit accountancy might have made those achievements look rather pointless), and the big mining projects in northern and western Australia were much puffed by industrialists and politicians. The 'resource booms' of the last decade of the second century of occupation were under way.

For the farmer and grazier, costs were now high. The land was netted with power and telephone lines as well as fences and roads and railways; life was easier. But not much of the original pastureland was now capable of carrying livestock at a profit without expensive superphosphates and introduced grasses. There were changes they had not expected; though the eucalypt forest had been cleared and the expenses of ringbarking, pulling and suckering were mostly in the past, trees left in pastures as shade or windbreaks were dying from unknown causes. Water resources were a further problem; most farms and stations put in big dams during the droughts in the 'sixties and late 'seventies but the flow in rivers was much reduced and most stock supplies as well as homestead supplies had to be pumped from below ground. This reliance on petroleum and electricity cost money and made life difficult during strikes and price rises. The injunction to 'get big or get out' was harder to ignore. The inland was little more populated than it had been fifty years before.

The grand irrigation schemes of the Murray–Darling basin which had indeed populated the area were showing signs of salinity and of water-table problems, and the water of the rivers was heavily polluted with a variety of different and unpalatable additions. The inhabitants of South Australia, relying on the Murray waters so heavily, complained of the greed and indifference of the other States which used the drainage basin. The neatly-squared blocks of orchards, vineyards and cropland which lined the river, men feared, might succumb to salinity within a few decades if nothing was done — but what could be done?

In 1980, Australian soil scientists issued a despondent report[6] on the state of the country's basic resource — soils. More than half of the land in production, whether of crops or pasture, needed urgent anti-erosion measures; some should be taken out of production altogether. The measures needed, however, would cost at least a billion dollars, a sum which made it most unlikely that they would ever be carried out. Governments intent on short-term popularity and with the next election in mind took little interest in such reports. Scientists blamed unwise timber-clearing, fallowing methods, overstocking of grazing land and mining operations for this deterioration — a process which could only worsen year by year.

Along the more heavily populated 'fertile arc' of coastlands between the Dividing Range and the littoral sea, many competing demands struggled for priority of occupation, almost always at the expense of the natural coastscape. Industrial development, tourist resorts, cities, sandmining, coalmining, 'infill' of coastal marshes, mangrove areas, tidal flats and estuary borders, garbage dumping, unwise building on dune areas, were among them.[9] Speculation in land, and almost total lack of planning controls, exacerbated the problem. Very little coastal parkland existed, and the pressures on what did exist increased yearly. Local councils tended to welcome almost any development which would increase population, whether or not it would also lessen the area's resources and its attractiveness. This rush to the coastal areas resulted by the 1980s in a much increased imbalance of population — observing Australia from a plane window it was possible to feel that it consisted mainly of an overcrowded and polluted littoral backed by almost empty spaces. The politicians of the century before, who had so confidently expected the growth of that industrious yeomanry which would people the inland would have been despondent at the sight, even though the expected invasions from China had not followed.

Often the industrial, tourist and suburban developments which fringed the eastern and southern coasts were set up at the expense of some of the most productive farmland on the best alluvial soils of the country. This had resulted in orchards and vegetable farms having to be moved to much less favourable sites, where water had to be provided at greater expense, fertilizer used more lavishly, and where distance from markets added to the costs of the farmer and the price to the consumer.

Though by the 1980s, the total population (this time the census did include Aborigines) was rising past the fifteen-million mark, the prospects for supporting many more millions were poor. Optimistic estimates

Top Cane farming making inroads into remaining stand of tropical rainforest, north of Cairns, Queensland.

Above A remnant of saltbush, which is palatable to sheep, contrasts with denuded grazed area at the fence line. (Photos: Colin Totterdell)

had put Australia's capacity for feeding its own people and those of other importing countries high. Average agricultural production could, it was said, provide food for from 52 to 70 million mouths and a continuing 40-odd per cent contribution by primary producers of food to the country's export earnings. But in 1980,[7] soil scientist Dr. Martin punctured this estimate and added 'There are signs that our productive capability is already crumbling.' Salination both in the irrigation areas and in dryland crop and pastureland added to the problems of erosion and ignorant soil management, and very little was known of Australian soils — and far less money was available for research, inventory and management planning than would be needed. This picture of a quickly deteriorating landscape would certainly not be radically improved by 1988.

As for the great plagues which were overtaking the country by the end of the nineteenth century, though the first intensity of their invasion had been overcome, some still troubled the land. The rabbits had left a changed landscape in the Darling country where some species of herbage and shrubs may have disappeared entirely, and in the Riverina those plants which rabbits preferred to eat gave way to such as they did not — white everlasting and cockspur. Says Eric Rolls[8] of the Riverina country after the rabbit invasion was subdued by myxomatosis: 'the quality of the feed was a long way below what it was, but at least something covered the ground'; and in other areas which had been heavily rabbit-infested, probably the same comment applies. Rabbits survive practically everywhere they had penetrated by the time of the myxomatosis introduction but are for the most part controlled by it so long as its vectors are present. Overstocking, particularly with sheep, was a factor as important as the rabbits themselves in the change in the landscapes of pastoral Australia, and has certainly done at least as much damage to the land; and continues so to do. The collapse of the prickly-pear which once covered such a vast area of Queensland was even more spectacular and seems permanent, though stray plants survive. Other weed introductions of the nineteenth century — thistles and burrs, lantana and blackberries — also survive and cost much to keep under control; eradication is a now abandoned hope.

Every drought which results in the importation of hay from one district to another provides a base from which weeds may take over new areas of land bared by overstocking; the long drought which ended the 'seventies of this century and persisted into the 'eighties was a potent factor in introducing such weeds as Paterson's Curse and various thistle species to coastal country in New South Wales, for the drought affected dairying and other farmland in areas seldom short of rainfall. There were complaints and accusations that the new arrivals of 'hobby farmers' and 'hippies' from the increasingly unpleasant cities were mistreating the country — it was at least as much misused by the previous occupiers, however.

The scene in the remaining forests, by the last decade of the second century, was much as before in that many species of commercially useful trees were still subject to over-exploitation, though this time under the auspices of foresters. The last virgin stand of tropical rainforest in the country, on Queensland's Mount Windsor Tableland, was being, so conservationists and ecologists reported, most damagingly logged. On the south coast of New South Wales and in Tasmania the effects of clear-felling for woodchip exports were becoming all too clear and woodchippers were moving into north-eastern Victoria to repeat the process. Private forests in that area had already been heavily exploited. The jarrah forests of southern W.A. were apparently destined for total destruction in the interests of bauxite mining and other forms of exploitation, added to new disease problems which were also appearing in the eucalypt forests of southern and eastern Australia. The clearing of native forests in the interest of plantations of introduced trees, mainly Pinus species, was a new factor whose final effects were still unkown but which added nothing to aesthetic perceptions of the landscape and was certainly unfavourable to wildlife survival.

Other factors new since the first centennial were myriad. In the northern part of the continent, apart from the vast new mining projects and the ephemeral new towns, the effects of uranium mining on waters, plants, animals and the Aborigines of the area were likely to be most sombre. Introduced animals had already changed the aspect of the north buffalo were destroying the productive swamplands by trampling, grazing and causing saltwater intrusions. They had already been responsible for the introduction of the cattle-tick which in the 1890s had reached right around the northern coasts beyond the Queensland border and caused immense losses in cattle, so that most cattle bred in tick-infested country now had strong infusions of resistant Asian strains. Wild pigs, feral cats, feral donkeys, camels, were among the variety of other introductions which were damaging either the land or its wildlife species.

By the last decade of the second century, then, an observer who had known the Australia of 1788 might not have recognised what he now saw, and might have grieved, in the words of a Queensland parliamentarian in 1901, that 'our Australia is . . . a very different place from what it was.'

At the beginning of the 'eighties there were a few hopeful signs. Conservation groups and paid-up membership of the Australian Conservation Foundation were numerous and rather more influential than they had been ten years earlier. Politicians intent on the short-term dollar had from time to time to take note of such pressures, and a number of reports had been commissioned and lodged whose recommendations, if they had been put into practice, would at least have made a creative response to the overall problems of the landscape and its inhabitants. Though the perception of Australia as a country to loot still ruled, and though its scars were many and deepening, there was a new realisation, among some, of the subtlety and uniqueness of its landscapes, and of the fact that they were also life-support systems whose changes could be as catastrophic to the dollar-nexus as to the non-human species of which they were made up. A new concern for what was now called 'the environment' had won certain small victories, on the Great Barrier Reef, on Fraser Island, on the sand-masses of the south Queensland coast, and in New South Wales, and had unsettled State governments over the exploitation of rainforests and the damming of the wild rivers of Tasmania. And such concepts as 'ecosystem', 'biosphere' and 'interface' were becoming more familiar. Though the resistance to these new concepts and pressures — which might stand in the way of speculators, developers and investors and hurt the wallets of taxpayers — is immense and though politicians therefore treat them with suspicion at best and contemptuous indifference at worst, the processes of education gave some hope of change.

But we are a nation of city-dwellers now, largely brainwashed by servitude to the interests of the corporations and the system which have inflicted such wounds on the Bicentennial Landscape. If it is to show signs that there is an intelligent determination to repair its wounds and attempt to live in harmony, instead of opposition, with its so urgent needs, there is terribly little time. (*Winter 1982*)

Bottom 'Alarmed settlers were complaining of the difficulty of clearing flourishing growths of prickly pear'.

Below 'Rabbits were eating grasses even down to the roots, killing trees and shrubs by nibbling off their bark'. (Photos: Australian Information Service)

The World Heritage List

Ralph Slatyer

Ralph Slatyer is Chairman of the World Heritage Committee.

In biblical times the seven wonders of the world reflected an awareness, by the scholars of that era, of outstanding examples of the world's cultural heritage. However, the boundaries of that particular world did not extend far beyond the eastern Mediterranean. The cultural heritage of other great civilizations was not represented, nor were outstanding examples of natural heritage.

Of these wonders, only one remains more or less intact, the Great Pyramid of Cheops in Egypt. It is now inscribed on the World Heritage List, a contemporary inventory of the wonders of the world which is being compiled progressively under the World Heritage Convention.

The World Heritage Convention exists to protect the world's cultural and natural heritage so that, unlike the seven wonders of the ancient world, properties on the World Heritage List can be conserved for all time. The Convention is, in many respects, a remarkable document since, while fully respecting national sovereignty, it recognises that cultural and natural properties of 'outstanding universal value' are of such importance to human civilization that it is incumbent on the international community to assist in their 'identification, protection, conservation, preservation and transmission to future generations'.

The Convention was adopted by the 17th General Conference of UNESCO in 1972. The United States, which played a leading role in the concept and development of the Convention, was the first signatory. Australia ratified the Convention in 1974. It came into force in 1975 when twenty countries had deposited an instrument of ratification or acceptance. Now there are more than sixty signatories with additional countries progressively adding their names.

Although the present participants are spread over all continents, there are still significant gaps in representation. Western Europe, north America and the African and Arab states are the best represented. The Asia-Pacific region is the least well represented with only five countries having so far ratified the Convention. Since these five comprise Pakistan, Nepal, India, Sri Lanka and Australia, there is no representation as yet from eastern and south-eastern Asia or the Pacific Islands.

The origins of the Convention lay in the growing awareness, towards the end of the 1960s, of the degree to which the rapid development of communications had increased awareness and interest, amongst the people of all nations, in cultures and regions different to their own. At the same time there was a growing awareness, which culminated in the U.N. Conference on the Human Environment, in Stockholm in 1972, of the degree to which rapid urban and rural development was threatening the cultural and natural heritage. Together, these elements led to an awareness that the deterioration or disappearance of the heritage in any part of the world constituted an impoverishment of the heritage of all humanity. With the support of UNESCO, and the encouragement of the international community, the Convention became a reality.

States Parties to the Convention commit themselves to help in the identification, protection, conservation and preservation of world heri-

The World Heritage List

The great temples of Rameses II at Abu Simbel. (Photo: Alexis Vorontzoff, UNESCO)

Properties included in the world heritage list

Contracting State having submitted the nomination of the property (in accordance with article 11 of the Convention), in italics.

Algeria: Al Qual'āof Beni Hammad.
Argentina: Los Glaciares National Park.
Australia: Kakadu National Park, Great Barrier Reef, Willandra Lakes Region.
Brazil: The historic town of Ouro Preto.
Bulgaria: Boyana Church, Madara Rider, Thracian tomb of Kazanlak, Rock-hewn churches of Ivanovo.
Canada: L'Anse aux Meadows National Historic Park, Nahanni National Park, Dinosaur Provincial Park, Burgess Shale Site, Anthony Island, Head-Smashed-In Bison Jump Complex.
Canada and United States of America: Kluane National Park/Wrangell-St. Elias National Monument.
Cyprus: Paphos.
Ecuador: Galapagos Islands, City of Quito.
Egypt: Memphis and its Necropolis, the Pyramid fields from Giza to Dahshur, Ancient Thebes with its Necropolis, The Nubian monuments from Abu Simbel to Philae, Islamic Cairo, Abu Mena.
Ethiopia: Simen National Park, Rock-hewn Churches, Lalibela, Fasil Ghebbi, Gondar Region, Lower Valley of the Awash, Tiya, Aksum, Lower Valley of the Omo.
Fed. Rep. of Germany: Aachen Cathedral, Speyer Cathedral, Wurzburg Residence with the Court Gardens and Residence Square.
France: Mont St. Michel and its Bay, Chartres Cathedral, Palace and Park of Versailles, Vezelay, Church and Hill, Decorated Grottoes of the Vezere Valley, Palace and Park of Fontainebleau, Chateau and Estate of Chambord, Amiens Cathedral, The Roman Theatre and its surroundings and the 'Triumphal Arch' of Orange, Roman and Romanesque Monuments of Arles, Cistercian Abbey of Fontenay.
Ghana: Forts and castles, Volta Greater Accra, Central and Western Region, Ashante Traditional Buildings.
Guatemala: Tikal National Park, Antigua Guatemala, Archaeological Park and Ruins of Quirigua.
Guinea: Nimba Strict Nature Reserve.
Hashemite Kingdom of Jordan: The Old City of Jerusalem and its Walls.
Honduras: Maya Site of Copan.
Iran: Tchogha Zanbil, Persepolis, Meidan-e Shah Esfahan.
Italy: Rock engravings in Valcamonic, The Historic Centre of Rome, The Church and Dominican Convent of Santa Maria delle Grazie with The Last Supper by Leonardo da Vinci.
Malta: Hal Saflieni Hypogeum, City of Valetta, Ggantija Temples.
Morocco: The Medina of Fez.
Nepal: Sagarmatha National Park, Kathmandu Valley.
Norway: Urnes Stave Church, Bryggen, Roros.
Pakistan: Archaeological ruins at Mohenjodaro, Taxila, Buddhist ruins of Takht-i-Bahi and neighbour city remains at Sahr-Bahlol, Historical Monuments of Thatt, Fort and Shalimar Gardens in Lahore.
Panama: The fortifications on the Caribbean side of Portobelo, San Lorenzo, Darien National Park.
Poland: Cracow's Historic Centre, Wieliczka Salt Mine, Auschwitz Concentration Camp, Bialowieza National Park, The Historic Centre of Warsaw.
Senegal: Island of Gorée, Niokolo-Koba National Park, Djoudj National Bird Sanctuary.
Syrian Arab Republic: Ancient City of Damascus, Ancient City of Bosra, Site of Palmyra.
Tanzania: Ngorongoro Conservation Area, Ruins of Kilwa Kisiwani and Ruins of Songo Mnara, Serengeti National Park.
Tunisia: Medina of Tunis, Site of Carthage, Amphitheatre of El Jem, Ichkeul National Park.
United States of America: Mesa Verde, Yellowstone National Park, Grand Canyon National Park, Everglades National Park, Independence Hall, Redwood National Park, Mammoth Cave National Park, Olympic National Park.
Yugoslavia: Old City of Dubrovnik, Stari Ras and Sopocani, Historical complex of Split with the Palace of Diocletian, Plitvice Lakes National Park, The Ohrid region with its cultural and historical aspects and its natural environment, Natural and Culturo-Historical Region of Kotor, Durmitor National Park.
Zaïre: Virunga National Park, Garamba National Park, Kahuzi-Biega National Park.

The World Heritage Emblem symbolizes the interdependence of cultural and natural properties: the central square is a form created by man and the circle represents nature, the two being intimately linked. The emblem is round, like the world, but at the same time it is a symbol of protection.   (© Unesco, 1978) tage sites and to refrain from taking any action which might directly or indirectly damage them. They recognise that the identification and protection of those parts of the heritage which are located on their own territories is primarily their responsibility, and agree that they will do all they can, with their own resources and with what international assistance they can obtain, to ensure adequate protection. They agree, amongst other things, to 'adopt a general policy which aims to give the cultural and natural heritage a function in the life of the community and to integrate the protection of that heritage into comprehensive planning programmes' and to 'take appropriate legal, scientific, technical, administrative and financial measures' necessary for its protection.

The Convention is administered by the World Heritage Committee, composed of twenty-one states, elected at a General Assembly of States every two years. Australia was elected at the first General Assembly in 1976 and has served continuously since then. I think it is fair to say that we have been one of the most active and constructive members of the Committee and have made useful contributions to the way in which the Committee works and the operational guidelines which it has progressively evolved; which, in turn, enable nominations for the World Heritage List to be examined in a professional and objective manner. Our standing in the Committee had been reflected in our election to vice-chairmanship of the Committee in 1980–81 and to the Chairmanship in 1981–82.

Properties which are nominated by States Parties to the World Heritage List are examined by the Committee each year. This process takes about ten months, the key step being the examination of each nomination by IUCN (the International Union for the Conservation of Nature and Natural Resources) for the natural properties and by ICOMOS (the International Council on Monuments and Sites) for the cultural properties. The IUCN and ICOMOS reports are discussed in detail by the Bureau of the Committee and then submitted to the full Committee for decision.

In order to qualify for the World Heritage List, a property must meet specific criteria of 'outstanding universal value' from either a cultural or natural point of view. If it is nominated as a cultural property, a monument, a group of buildings, or a cultural site, must:

- represent a unique artistic achievement, a masterpiece of the creative genius; or
- have exerted great influence, over a span of time or within a cultural area of the world, on developments in architecture, monumental arts or town-planning and landscaping; or
- bear a unique testimony to a civilization which has disappeared; or
- be an outstanding example of a type of structure which illustrates a significant stage in history; or
- be an outstanding example of a traditional human settlement which is representative of a culture and which has become vulnerable under the impact of irreversible change; or
- be directly or tangibly associated with events or with ideas or beliefs of outstanding universal significance (the Committee considered that this final criterion should justify inclusion in the List only in exceptional circumstances or in conjunction with other criteria).

If a property is nominated in terms of its natural properties, it must:
- be an outstanding example of major stages of the earth's evolutionary history;
- be an outstanding example of significant ongoing geological processes, biological evolution and man's interaction with his natural environment (as distinct from the first criterion this focuses upon on-going processes in the development of communities of plants and animals, landforms and marine and fresh water bodies); or
- contain superlative natural phenomena, formations or features of exceptional natural beauty, such as superlative examples of the most

Australia in Trust

Far right The Hook and Hardy Reef is part of the Great Barrier Reef, the most extensive reef system in the world. (Photo: Great Barrier Reef Marine Park Authority)

Right The square of San Francisco in the old city of Quito, Ecuador. (Photo: A. Abbé, UNESCO)

Below Dr Rhys Jones indicates an Aboriginal hearth on the lunette at Lake Mungo in the Willandra Lakes Region of western New South Wales. (Photo: Australian Information Service)

- important ecosystems, natural features, spectacles presented by great concentrations of animals, sweeping vistas covered by natural vegetation and exceptional combinations of natural and cultural elements; or
- contain the most important and significant natural habitats where threatened species of animals or plants of outstanding universal value from the point of view of science or conservation still survive.

So far one hundred and twelve properties have been inscribed on the World Heritage List. These comprise seventy-eight cultural and thirty-four natural properties, although some properties have met both natural and cultural criteria for inclusion. Twenty-seven of these properties were added to the list at the 1981 session of the World Heritage Committee which was held in Sydney in October at the invitation of the Australian Government.

This session, which was opened by the Prime Minister, Mr. Fraser, was of particular interest to Australia because three Australian nominations — The Great Barrier Reef, Kakadu National Park and the Willandra Lakes Region — were under consideration by the Committee. All three were adopted and Committee members subsequently had the opportunity of seeing parts of the Great Barrier Reef at first hand.

The World Heritage List now contains many of the most famous elements of the world's cultural and natural heritage. Included amongst them are, for example, Sagamartha National Park, which includes Mount Everest; the Egyptian pyramids and the Nubian monuments from Abu Simbel to Philae; Serengeti National Park; the old city of Jerusalem and its walls; the Galapagos Islands; and the Palace of Versailles. In addition, there are properties less well known to Europeans or Australians but undoubtedly of outstanding universal value. They include outstanding Islamic monuments such as the Medina of Tunis and the archaeological ruins of Mohenjadaro in Pakistan; outstanding cultural sites in Latin America such as the old city of Quito in Ecuador and the Mayan monuments of Tikal in Guatemala. The full list, as at August 1982, is given in the box.

The three Australian properties which were accepted in 1981 were favourably reviewed by IUCN and ICOMOS and have earned their place in the List in every sense. In fact both Kakadu National Park and the Willandra Lakes Region had the unusual distinction of being recommended for both their cultural and natural significance.

The Great Barrier Reef needs little introduction to most Australians. It is the world's most extensive stretch of coral reefs, comprising some two thousand five hundred individual reefs of all sizes and shapes, providing the most spectacular marine scenery on earth. It is also probably the world's richest area of faunal diversity, with over one thousand five hundred species of fish, about four hundred species of coral, four thousand species of molluscs and two hundred and forty species of birds, plus a great diversity of sponges, anemones, marine worms, crustaceans and other groups of animals and plants. It includes major feeding grounds for the endangered Dugong, and nesting grounds for two endangered species of marine turtles — the green and loggerhead turtles. The IUCN review of the Great Barrier Reef stated that it met *all* the natural criteria for inclusion of a property on the World Heritage List, and concluded that 'if only one coral reef site in the world were to be chosen for the World Heritage List, the Great Barrier Reef is that site'.

Kakadu National Park, in addition to its outstanding natural features, which alone would have warranted its inclusion on the World Heritage List, was also recommended by ICOMOS because of its cultural value. The ICOMOS evaluation emphasized, in particular, the Aboriginal rock paintings which cover a chronological span from 20,000 years ago to contemporary time. ICOMOS commented that these paintings 'constitute a fund of documentary evidence of primordial importance and a source which is unique. They serve as a source of information on the primal resources, the hunting and fishing activities, the social structure and the ritual ceremonies of the Aboriginal population which have succeeded one another on the site of Kakadu. They bear witness to vanished species, such as the Tasmanian wolf, and allow one to follow, in the details of equipment and of costume, the modifications brought to bear on traditional life by the contacts which were established with the Macassan fishermen from the 16th century, and then with the Europeans'.

The Willandra Lakes Region also satisfied both natural and cultural criteria. ICOMOS commented that the region had a world wide importance owing to the 'abundance of vestiges of very early human occupation'. The evaluation emphasizes the importance of both the oldest dated evidence, from about 40,000 years ago, and also the cremation graves which showed a carbon date of 26,000 years, being the oldest examples discovered anywhere in the world. ICOMOS considered the region of comparable importance to the Omo and Awash valleys in Ethiopia.

At the end of 1981 Australia submitted two additional nominations — the Western Tasmanian Wilderness National Parks and Lord Howe Island. The former area was the subject of the recent controversy concerning the proposal to dam the Franklin River. Both sites are now inscribed on the List. (*Summer 1982*)

Lord Howe Island

Chris Murray

Chris Murray is the Vice-President of the Lord Howe Island Preservation Society.

While Captain Arthur Phillip was struggling with his infant colony at Port Jackson, he despatched, in February 1788, one of his trustiest vessels, the armed tender *Supply*, to sail to Norfolk Island and there found a second colony. It was en route to Norfolk Island that *Supply*'s Captain, Lieutenant Henry Lidgbird Ball, discovered Lord Howe Island and the adjacent pinnacle that still bears his name, Ball's Pyramid.

The *Supply* did not anchor at Lord Howe until the return journey when Ball and his crew found an abundance of turtle and a truly amazing variety of birdlife. Unaccustomed to predators, the birds showed no fear of the ship's crew. Arthur Bowes, surgeon of the *Lady Penrhyn*, which called shortly after, exclaimed in his diary 'When I was in the woods amongst the Birds I cd. not help picturing to myself the Golden Age as described by Ovid . . . .'

Alas, the 'Golden Age' soon passed in this Arcadian paradise. The first men intruded on the Island as hunters and gatherers in a style reminiscent of much human exploitation since the dawn of prehistory. The larger ground birds were soon hunted to extinction by hungry foragers from the First Fleet and by whalers, who came to know the Island as a place for fresh water and game.

These periodic forays by visiting ships were the main human impact on the Island until 1834 when three New Zealanders, their Maori wives, and two boys, settled on the flats behind Old Settlement Beach. They established gardens and made an industry from the slaughter of mutton birds for their feathers.

In 1841 they were bought out by Owen Poole, a retired officer of the Bombay Establishment, and Richard Dawson, one of Sydney's first ironfounders. Poole subsequently sold half his share to Dr. Foulis, who brought his family to the Island in 1844. However this impressive triumvirate of gentlemen squatters persisted for only a few years. When, in 1847, the Colonial Government refused to cede them a leasehold on the Island, they departed, leaving their more humble retainers, the Andrews, Wrights and Moseleys to farm the land. As time passed these families were joined by others, including stalwart mariners like Thomas Nichols and Nathan Thompson. (Thompson arrived on a whaler in 1853 with three native women from the Gilbert Islands, one of whom was a princess escaping from an arranged marriage.)

So the real settlement of Lord Howe Island was brought about by humble men and women, the progenitors of a hardy race of Islanders, who were able to turn an ingenious hand to whatever bounty the gods of opportunity offered.

Throughout the 1850s and 60s they bartered 'livestock, Fish, Potatoes and other Vegetables, Slops etc' with the passing whalers of the 'Middle Grounds'. Whaling captains, glad of the fresh provender, traded all manner of goods with the thrifty Islanders including, on one occasion, even the ship's cannon.

Whaling collapsed in the 1870s and there were some lean times on the Island, including a period of three years when no ships called at all. In spite of difficulties, the Islanders kept up an intermittent trade with the mainland in produce, mostly onions and oranges. The onion trade literally grew from a single bulb which was washed up on an Island beach, while the first orange trees sprouted from some pips which arrived providentially on a passing whaler from Tahiti.

It was not until the final decades of the nineteenth century that the fortunes of the Island people once more flourished, this time through the sale of seed from the endemic *Howea* palm. (These 'Kentia' palms are still considered by nurserymen the world over to be one of the most classically beautiful of all indoor palms.) The palmseed industry, and the fortunes of the Island generally, were cemented by a regular Burns Philp steamship service from 1893.

Unfortunately the trade in palmseed became so profitable that disagreements about pricing broke out between some Islanders and the mainland nursery companies they supplied. The State Government intervened, sponsoring two commissions of inquiry, the second in 1911. As a consequence the palmseed industry was reorganised with all Islanders holding equal shares in the trade. Thenceforth the whole community worked co-operatively to harvest, pack and ship the seed in a manner which has been described as 'the perfect form of socialism'.

The Royal Commission also gave the Island its first regular form of local government, a Board of Control, which, one suspects, was not always to the liking of a community that had largely been free to organise its own affairs until this time.

Alas, the fortunes of the palmseed industry staggered under the blow of a rat plague in 1918, when the ship *Makambo* ran aground near Ned's Beach. The rats escaped from cargo removed from the vessel for safe storage, and soon began to prey on palmseeds, birds' eggs and the Islanders' homes. The seed harvests declined drastically and four more of the Island's endemic bird species suffered extinction. The Islanders battled manfully against the rats, in one year alone catching 20,000 of them.

In spite of these efforts the seed industry never quite regained its former dominance, and some Island families began to take tourists who were travelling on the Burns Philp ships. By the time of the second World War guest-houses were a well established part of the Island economy.

During the difficult years of the war the tourist trade disappeared, but the Islanders survived by growing vegetable and flower seed for mainland seed companies like Yates and Rumseys. After the war tourism returned with a flourish, this time through the means of converted Sunderland and Sandringham flying boats. Operating out of Sydney Harbour, the seaplanes often flew as many as five times a week in the summer months, though the service was less regular in the winter.

In 1974 the State and Federal Governments funded the construction of an all weather airstrip and today the Island is serviced by small commuter aircraft from Sydney, Brisbane and Port Macquarie. Tourism continues as the Island's main economic prop, though Government employment and the seed industry are also important.

The airstrip era has ushered in many complex social and political changes, some of which will be mentioned later on. Until this time the Island community retained a very definite sense of its own identity. While politically tied to New South Wales, and while, also, the Islanders' sympathies were with 'the mainland' (the names on the Island cenotaph attest to this), the community always considered itself to have a separate identity. During the heyday of the seed industry Islanders alone were entitled to shares in the industry and their status was legitimised in the *Lord Howe Island Act* of 1953. Only 'the holder . . . of a permissive occupancy of part of the Island from the Board of Control . . . (on) . . . the 4th of February, 1913 . . . or the issue of such person . . .' was entitled to apply for 'vacant Crown lands for the purpose of residence'. This recognition of one's land rights in connection with one's ancestry must be unique in the

Top The lagoon from Cobby's Corner. (Photo: M. Morris)

Far left Big Mountain Palm, *Hedyscepe canterburyana*. (Photo: B. Miller, N.P.W.S.)

Middle left The woodhen, *Tricholimnas sylvestris*. (Photo: B. Miller, N.P.W.S.)

Bottom left Pumpkin tree, *Negria rhabdothamnoides*. (Photo: B. Miller, N.P.W.S.)

history of Australia, at least until recent times when Aboriginal land rights are at last receiving some attention.

The legal definition of Islander status was substantially altered in the *Lord Howe Island Amendment Act* (1981), a ten year residency period being substituted for lineal descent. However the concept of 'Islander preference' is still retained in land allocation and as a qualification for election to the Lord Howe Island Board. While the new definition has allowed a small number of newcomers to apply for Islander status, the majority of Islanders can still trace their ancestry to the early settlers.

Sadly, it is the Island identity, in a psychological sense, that has been seriously eroded. The halcyon days of community co-operation in the palmseed industry have long gone, and in recent years there has been a wholesale sellout to the soft soap of suburbia — motor cars, video T.V.,

View from the summit of Mount Gower. (Photo: M. Morris)

and telephones. It is difficult to retain a separate identity when one is completely submerged in the lifestyle and paraphernalia of western society.

Predictably the Island's material affluence has brought with it a fifth column of social disruption in the form of increased delinquency, alcoholism and drug abuse, particularly amongst the younger Islanders. A once clear headed community now finds itself confronted with a bewildering array of social problems which simply did not exist in the 'old days'. In this respect the Island has much in common with other western communities in modern times.

While the Island people may be losing their sense of cultural identity, the staggering natural beauty of the Island remains. Seeing it for the first time is an experience one can never forget: to the south the towering, volcanic ramparts of Mount Gower and Mount Lidgbird; in the north the softly undulating line of the Malabar Hills; and between these the green, verdant forests and fields of the lowlands. Breakers swell and curl on a coral reef, fringing with foam the boundaries of an opalescent lagoon. Sixteen kilometres to the south juts the lone, Gothic spire of Ball's Pyramid. Thereafter, as far as the eye can see, lies the unbroken blue of the Pacific Ocean.

The Island's precious natural assets read like a 'who's who' of natural history: most southerly coral reef in the hemisphere; one third of the plant species are found nowhere else in the world; home of the Lord Howe Island Woodhen, one of the rarest birds in the world; only known nesting ground of the Providence Petrel; fossil site of the extinct, megalithic horned turtle. A conscious appreciation of the Island's uniqueness as a capsule of natural history pervades all scientific reports from the very earliest to the most recent.

Geologists tell us that the Island came into being nearly seven million years ago, the result of a volcanic upheaval on the Lord Howe Rise, a massive chain of seamounts extending north/south for well over 2,000 km. A second period of volcanism 6.3 million years ago completed the volcanic structure of the Island which was, then, a large shield volcano some 50 kilometres long, 16 kilometres wide and perhaps 1,700 metres in height. Today the Island is a slim crescent, 12 kilometres in length and only two kilometres at its widest. Mount Gower, the tallest peak, stands 866 metres above the sea. Clearly the Island, as we know it today, is only an eroded if spectacular remnant of its former self.

The present lowland areas of the Island were built from basalt alluvium washed from the hills and mountains and by wind-blown coral sand deposited during the last ice age. The coraline sands were cemented into calcarenite sandstone which helped to fossilize remains of *Meiolania platyceps*, the curious horned turtle which spent most of its life on dry land. Carbon dating suggests that Meiolania became extinct more than 20,000 years ago, though the broader questions of why it became extinct, or even how it came to inhabit Lord Howe in the first place, remain a complete mystery.

Reef building is now largely confined to the barrier reef which forms the present lagoon on the western side of the Island, though reefs may have been more extensive when sea levels were lower in the past. The outstanding fact about the Island's marine environment is the fluctuation in water temperature brought about by the intrusion of both warm, tropical currents and cool, temperate currents. Thus marine animals and plants, from tiny reef builders to large fish, are an unusual mix of tropical and temperate forms. The reef itself is a rare transitional form containing both coral and algal elements, the coral being the most southerly in the hemisphere. About four per cent of the fish species are unique to the Island and adjacent waters of Norfolk Island and Middleton Reef.

The Island's flora is an equally intriguing mixture containing 177 plant species whose affinities can be traced to three widely differing areas — New Zealand, eastern Australia and New Caledonia. There are, in addition, a few surprising anomalies like the lovely Wedding Lily (*Dietes robinsoniana*), whose closest relatives are to be found in South Africa.

Some 57 plant species are unique to the Island, including three endemic genera of palms. One of these, *Howea forsterana*, has become the doyen of indoor palms, with nurserymen from Europe, America and Australia prepared to bid up to $500 per bushel for seed collected from the Island. Other unusual species include the Pumpkin Tree (*Negria rabdothamnoides*) and the Giant Heath (*Dracophyllum fitzgeraldii*), both amongst the only tree-sized members of their families in the world.

An Australian Museum environmental survey has identified 16 different types of plant communities varying from stunted scrub and grassland on more exposed sites to dense, misty rainforest on the summits of Mount Gower and Mount Lidgbird. In itself such variety is extraordinary when one considers that the Island is a mere 1,455 hectares in area.

One theory on Lord Howe's unusual botanical inheritance suggests the Island is actually a living fossil remnant of a habitat which, in earlier and moister times, existed over wide areas of eastern Australia and the South Pacific. While such theories are difficult to prove, the fact remains that we are confronted with a fascinating ecosystem where many unique plants and plant communities have evolved or been preserved.

The only mammal native to the Island is a small bat, and terrestrial reptiles are meagrely represented by two lizards, a gecko and a skink (now largely confined to the offshore islets because of predation by rats). There are no snakes or venomous spiders.

This rather sparse animal population is more than balanced by prolific numbers of land and sea birds which occupy almost every available niche in the environment. Tropic Birds and Black-winged Petrels nest on the craggy headlands and cliffs; shearwaters burrow underground; waders and cranes scour the Island's foreshores, pastures and swamps; Masked Boobies, noddies and terns find sanctuary in the offshore islets; blackbirds, songthrushes and doves spend much of their time foraging on the forest floor; silvereyes, kingfishers, Golden Whistlers and currawongs live in the canopies and sub-canopies of the Island forests; last, but not least, the Woodhen and Providence Petrel cling tenaciously to the mountain summits.

Of the seabirds, some twelve species nest regularly on the main Island and offshore islets. The colonies are reasonably accessible and the birds show almost no fear of human intrusion.

The original complement of fifteen indigenous land birds has been sadly depleted with only five species remaining — the Golden Whistler, Sacred Kingfisher, Lord Howe Island Silvereye, Pied Currawong and Woodhen. However the gaps have been filled by both migrant and introduced species like blackbirds, songthrushes and mudlarks. There are, in addition, some 79 species of land and sea birds known to visit the Island but not to nest there.

Of particular interest is the fate of the Lord Howe Island Woodhen, often described as 'the rarest bird in the world'. The population of this flightless rail was so badly decimated by the impact of human settlement (particularly the introduction of cats, rats and pigs) that it survived, until recent years, in only two tiny enclaves on Mount Gower and Mount Lidgbird. These precarious colonies, taken together, numbered no more than 20 to 30 individuals.

Commencing in 1969, John Disney of the Australian Museum began to monitor and carry out basic research into the woodhen population. Year after year John and his associates visited the Island and made the arduous trek to the summit of Mount Gower to band woodhens and observe them in their last remaining habitat. Then, in 1978, the National

Top Black-winged petrel, *Pterodroma nigripennis*.

Above White tern, *Gygis alba*. (Photo: B. Miller, N.P.W.S.)

Parks and Wildlife Foundation came to the rescue with some $280,000 for a captive breeding project, this money having been generously donated by many Australians for the purpose of preserving endangered species. A further two years of intensive research was undertaken by Dr. Ben Miller of the National Parks and Wildlife Service. His aim was to replicate, as closely as possible, the woodhen's natural environment within a captive breeding enclosure on the lowlands.

Finally, in May 1980, three pairs of woodhen were whisked by helicopter from the summit of Mount Gower to their new home in Steven's Reserve. At that point their welfare became the responsibility of New Zealand aviculturist Glenn Fraser, who has so far managed to breed some 75 woodhen chicks within three breeding seasons, making this one of the most successful captive breeding projects anywhere in the world.

One would like to feel that the saga of the woodhen, with all its planning, expertise and devotion given for the preservation of one fragile species, is a good omen for the future of the Island itself. For unless as much care and attention are given to the fragile qualities which make Lord Howe Island unique, those qualities are likewise in danger of becoming extinct.

In recent years there has been a growing recognition of the need to preserve Lord Howe Island, not only its lovely natural landscape, but also, to some extent, the human heritage and unhurried lifestyle of the people who live there.

Concern has been expressed in many ways. These include the compilation of some excellent reports about the Island, including a major environmental survey by scientists of the Australian Museum and two town and country planning reports by eminent men in this field. Conservation bodies like the Total Environment Centre and Nature Conservation Council have promoted the conservation ethic by actively lobbying government decision makers. Their efforts have, in turn, been supported by many fine articles about the Island published in conservation journals and in the popular media. Finally, one must mention the personal efforts of men like Jim Brown and Vincent Serventy who have worked tirelessly to promote the Island's conservation.

The expression of so much articulate concern has influenced policy-making on all levels of government. In January 1981 the Wran State Government proclaimed the *Lord Howe Island Amendment Act* which marked a definite move in the direction of conservation:

1) The Lord Howe Island Board was reconstituted to include a nominee of the Minister for Planning and Environment
2) A 'Permanent Park Preserve' (revocable only by an Act of Parliament) was proclaimed over the untouched natural areas at the north and south ends of the Island
3) The Board was required to adopt a Plan of Management for the Permanent Park Preserve and to finalise a planning scheme for the settled portion of the Island
4) The Board was defined as a local council under the Planning and Environment Act and was henceforth required to comply with the provisions of that Act.

The Wran Government, in conjunction with the Fraser Federal Government, also moved to nominate the Island for inclusion on the World Heritage List. The nomination was accepted in December 1982 after a detailed submission had been presented to UNESCO in Paris by representatives of the Australian Heritage Commission. World Heritage listing marks the Island as being 'of outstanding universal value from the aesthetic or scientific point of view' and places it under the protection of the World Heritage Convention.

With such an impressive list of environmental safeguards it would seem that the future of conservation on the Island is assured. However, as

Top The Admiralty Islets, a nesting ground for many seabirds. (Photo: M. Morris)

Far right Wedding lily, *Dietes robinsoniana*. (Photo: B. Miller, N.P.W.S.)

Right Bearded mullet, *Parupeneus fraterculus*. (Photo: N. Whitfield)

observers of the Franklin Dam controversy in south-west Tasmania would agree, environmental safeguards are only as effective as the governing powers wish them to be. In the case of the Island, all the administration and many crucial policy decisions are the responsibility of the Lord Howe Island Board. Unfortunately, the Board not only faces a plethora of environmental problems, but operates within some severe limitations.

The Island's environmental problems are legion: polluted ground water supplies; garbage disposal problems; large areas of natural bush and foreshore threatened by grass and noxious weeds; feral animal populations that are hard to control; and pressure to allocate more and more land for residential and business purposes.

Against these problems the Board labours with a shortage of finance and lack of expertise. Finance is probably the major difficulty, the Island community of some 280 people being much too small to provide a useful amount of local government revenue. Traditionally the Board has fallen back on the palmseed industry for income. However, like the economy in general, this industry has taken a downturn, and the Board's newest scheme to establish an export nursery on the Island has not, so far, met with financial success. At the time of writing, the Board had practically no reserve funds for general administration let alone for conservation purposes.

Thus the Board is able to employ only one ranger for one or two days per week to take care of that 70 per cent of the Island which falls within the 'Permanent Park Preserve'. The ranger's time is overtaxed and many basic programmes which are carried out in national parks are simply non-existent. There is no monitoring of bird and wildlife populations, practically no noxious weed control on reserve lands, no guided tours or lectures for visitors to the Island and no field assistance for scientists making field trips to the Island. This is a sad state of affairs, when one considers that the Permanent Park Preserve was trumpeted by the State Government as being 'even better than a national park'. The Board's financial problems were accurately predicted by conservation bodies who have consistently lobbied the State Government for direct National Parks and Wildlife Service administration of the Island's reserves.

Shortage of funds has also curtailed the Board's ability to enforce regulations. While residents receive a continuous stream of notices from the Board about control of noxious weeds, stockproof fencing, littering, speeding motor vehicles and other matters, these edicts are hardly ever enforced. Indeed the Board's 'paper bureaucracy' has become something of a joke on the Island.

The Board is equally hampered in its efforts by a lack of expertise to deal with environmental problems. Many excellent reports have been compiled by environmental consultants, but the recommendations are not always clearly understood or acted upon. To overcome this problem, a scientific advisory committee was formed in 1981 with members from the Australian Museum, the National Parks and Wildlife Service, the National Herbarium and other bodies. Its aim was to provide the Island with advice on environmental matters but, for some reason, it met only once and was never reconvened by the Board.

The Amendment Bill of 1981 gave the Island people a majority of three elected representatives on the five man Board. Prior to this they had only one representative. For the local population increased representation has many benefits, but it has also given the Board an Achilles heel in terms of vulnerability to local pressure. The Island Board members are only human, and they cannot be blamed for keeping a weather eye on the feelings of the electorate, even if these run counter to the long term interests of the Island.

Two really vital areas where the Board is being subjected to some pressure include the demand for more residential land and a demand to increase the present tourist ceiling. The heavy demand for land has led to the expedient of approving condensed, suburban type housing in open paddocks. This new development stands in poor contrast to the older, more natural style of spaced leaseholds with intervening greenbelts and trees.

Pressure to lift the tourist ceiling from its present level of 400 licensed beds is strong, in spite of the fact that the real tourist intake is already well above the official limit. This situation occurs because Island families host large numbers of friends and relatives in their own homes, particularly during the summer holiday period. Family visitors are not considered in the 400 bed limit even though they make the same demands upon the Island as other visitors. With problems of ground water pollution and garbage disposal already serious, a further increase in the tourist ceiling is simply not justified.

It is a matter for conjecture just how the Island will be affected by these pressures in the long term. The Board has wisely placed a freeze on major development while it considers the merits of two alternative planning proposals — one by Nigel Ashton of the State Planning Authority and one by Professor John Toon of Sydney University. The implications of each plan are too detailed to be explained here. It is sufficient to say that both are comprehensive planning strategies which contain some stringent and some open-ended recommendations. The balance between conservation on one hand, and exploitation on the other, will depend on which elements the Board selects for the final plan, and in this area it is very vulnerable to the feelings of the local electorate.

Motor vehicle numbers are yet another electorally sensitive issue. There is a strong consensus among environmental consultants, visitors and even some Islanders that the present burgeoning vehicle population should be subject to some sort of limitation. Island roads, once the preserve of walkers and cyclists, are now dominated by speeding cars and motor cycles. Although the Board has been given complete regulatory power to control motor vehicle numbers, it has not yet found the courage to do so.

In all fairness to the Island's administration it must be stressed that the Lord Howe Island Board has made some difficult decisions in the interests of the environment. These include banning the importation of domestic cats, introduction of the Dog Act, and the current freeze on major development. The Board's difficulties are not so much the result of its members' attitudes (most of whom express a commitment to conservation) but to the limitations of the Island itself. One simply cannot expect the finance and expertise necessary to administer a World Heritage area to come from a small community of 280 people. Indeed one feels that the Board may well be overwhelmed by the magnitude of the task because it has too few resources to pit against too many environmental problems.

In a more positive vein, many difficulties could be dealt with by increased assistance and funding from outside. It would be good to see a scientific advisory committee reconvened, the National Parks and Wildlife Service more closely involved with the Island, government funding earmarked for projects like noxious weed control and feral animal eradication. Much more attention could also be given to the marine resources of the Island which still have only a vague form of legislative protection and would benefit by being administered as a marine park.

Ideally, one would like to see a vision of the future involving co-operation on all levels from the community up, resulting in a profound and comprehensive care for the Island as part of our Australian and World Heritage. It might well be that such a noble cause would give the Island people a renewed sense of purpose. If, however, the essential involvement and co-operation are not forthcoming, it is still quite possible that the environment will be degraded in spite of all the existing safeguards.

We must resolve now that Lord Howe Island will not become another silent victim of man's inability to sustain higher aesthetic and moral aims against narrow, parochial self-interests. (*Summer 1983*)

Kakadu — man and landscape

Text and photographs by Allan M. Fox

Allan Fox is a consultant in environmental management and is also a well-known landscape photographer.

The Aborigines

Left Sorcery Rock, near Obiri.
Bottom Twin Falls in the wet season.
Below Mount Brockman. (Photos: Allan Fox)

The date is November 21st, 1845 and Ludwig Leichhardt has just descended with some difficulty into Deaf Adder Gorge by way of 'rocky creeks, between loose blocks off which our feet were constantly slipping'. Leichhardt had arrived in the country of the Bardmardi clan. To this first European, the crossing of the rocky gullies and sandstone plains of the Arnhemland Plateau had been full of interest albeit static . . . 'We had observed a great number of shrubs, amongst which a species of *Pleurandra*, a dwarf *Calythrix*, a prostrate woolly *Grevillea*, and a red *Melaleuca*, were the most interesting.

'The melodious whistle of a bird was frequently heard in the most rocky and wretched spots of the tableland . . . (it) was very pleasing, and frequently the only relief while passing through the most perplexing country.'

Perplexing indeed! For our first European observer, there were many new and interesting things to list, but the observations were as static as a still picture and considerably less detailed having passed through perceptual filters developed in nineteenth century Europe. Like many who followed, Leichhardt was 'discovering' Australia, and ignoring the fact that the place had been occupied for forty thousand or more years; 2,000 generations at least by Aboriginal Australians.

Throughout the build-up to the wet season Leichhardt and his party would have traversed what is now Kakadu National Park from its southeast corner to the mouth of the East Alligator River — western civilization had arrived.

The story of Kakadu begins more than 2,000 million years ago, when mud, sands and boulders were swept by river systems into a sea . . . there were times too when volcanoes ejected vast quantities of ash and lava onto this great bed of foundation material. The time scale is enormous, long before any animal life was abroad. These ancient rocks were bent, warped and folded. Injections of igneous materials and mineral-charged solutions worked through the mass via the faults and fractures. These areas subsided to form a vast sunken geosyncline. Into this trough from about 1,600 million years ago sediments, kilometres deep were dumped. The layers were mixed; sometimes powerful streams carried large boulders, in less violent times, sand and mud. At times too, volcanoes opened up, pouring lava across the landscape; at other times ash and breccia (broken rock). There were times too when lava failed to reach the surface but ran between the layers in sills. As the topography, through erosion and deposition, became low and subdued, the rivers lost their power. Following a quiet period stresses working on the earth caused the whole area to uplift with a gentle slope to the north. These later, nearly horizontal lying strata, took on the immense grid of fractures in the older underlying rocks. Erosion continued with the streams following these joints and subsequent erosion has produced the landscapes of today ... the Arnhem Plateau, more than 1,000 km of escarpment and numerous outliers lying like mountainous islands on the deeply weathered lowlands and broad floodplains and coastal plain, each area with its particular assemblage of plants and animals ... grasslands, savannah, woodlands, forests, monsoon forests, sedgelands and even pseudo-desert heathland on the sandstone tops.

That is the way Kakadu looked to Leichhardt, but the Aborigines certainly didn't arrive finding the scene that way. It is most likely that the ancestor Aborigines arrived at a time when a glacial period was in full swing. The sea level had dropped some 100 metres. The last time the levels were as low as that was from 15,000 to 20,000 years ago, but at that time there were people grinding their tools near the caverns by the East Alligator River and Aborigines had been camping by Lake Mungo for at least 10,000–15,000 years. It is probable that the likely arrival period was during the low level previous to the last, between 55,000 and 50,000 years ago.

Even with these low sea levels the trip to Australia required a substantial voyage of forty or fifty kilometres, to the combined North Australian –New Guinean coast 300–500 kilometres north-west of the present coast.

Climates too during these glacial periods would have been colder and drier. The source of moisture was further away and there was some 70% less evaporation from the cooler oceans. The formation of tropical cyclones was markedly reduced and the high pressure belt was more persistent over the continental desert. During the glacial periods the deserts marched north, but by 9,000 years ago the situation had reversed with rainforests and monsoon forests covering their maximum area.

The landscape/ecosystem image has been in a constant state of flux ... estuaries became fresh water rivers ... became entrenched rivers ... became broad embayments ... became saline wetlands ... became freshwater swamps ... became seasonally dry floodplains. In terms of vegetation: mangrove forests fronting samphire flats became *Melaleuca/ Barringtonia/Pandanus* forests fronting grassed and savannah plains ... became marine mudflat areas ... became mangrove/samphire ... became sedgelands/grasslands and if the process continues, savannah. With each of these merging changes in habitat so follows a changing fauna. Change is the essence of Kakadu and to survive for 2,000 or more generations the Aboriginal people were required to be highly adaptable. How did the Aborigines perceive all this?

Traditionally, the Aborigines believe that their great ancestors came ashore on a northern beach and wandered the land creating its physical features and the creatures, including man. Incidents in their wanderings and subsequent spirit activity have given rise to the landscape, but more than that, the spirits established for all time patterns of behaviour governing the relationships between man and nature and above all, with the spirit ancestors themselves.

The first Chairman of the Northern Land Council, Silas Roberts, put it this way:

Aboriginals have a special connection with everything that is natural. Aboriginals see themselves as part of nature. We see all things natural as part of us. All the things on Earth we see as part human. This is told through the idea of dreaming. By dreaming we mean the belief that long ago, these creatures started human society. These creatures, these great creatures are just as much alive today as they were in the beginning. They are everlasting and will never die. They are always part of the land and nature as we are. Our connection to all things natural is spiritual.

Add to this deep attachment to all things natural, 40,000 years or more of experience in these ecosystems, then we will begin to understand how rich must be their resource-oriented knowledge and why our perceptions of the Kakadu environment are so simplistic by comparison. The Aboriginal landscape is fully personalised. But how is such a store of tradition passed on without a written language?

It happens primarily because life and learning are synonymous. Even a cursory look at the Bardmardi year will illustrate the point. Tradition is continuously demonstrated and practised as a staged learning process culminating at its richest point at the moment of death. Transmission is by way of myth, stories, field experiences, play, song cycles, dances, pattern and design of artefacts and the ritual of ceremonies.

Leichhardt arrived during the stormy build-up to a fairly normal wet season and no doubt wondered where all the Aborigines were because Deaf Adder Valley seemed such a paradise . . . large barramundi, whistle ducks, pigeons and freshwater crocodiles were in abundance. This was the season of *Gunumeleng*, the pre-monsoon storms when the early rains had greened the land and the billabongs were filling. Magpie geese were fat and the people felt a longing for goose. Leaving their cavern shelter the families moved out onto the South Alligator floodplain. Here with the local people they selected a belt of paper-bark trees which lay across the flight paths of the heavy, low flying geese. A stack of waterlogged sticks was placed on a platform built high in the canopy. As the skeins of geese honked across, the sticks were hurled into the mass bringing down numbers which were soon dispatched by the children. It was out on the plain where Leichhardt met most of the Bardmardi. Here too he heard Aborigines using 'Malay' (Macassan). Hundreds of people gathered from near and far for the enactment of major ceremonies.

With the end of the ceremonies the families cut suitable bamboo stems for spear shafts and on the way back to Deaf Adder walked via an outcrop of Oenpelli Dolerite selecting stone blanks for the manufacture of tools. All the while the children and young people were enriching their experience watching closely and copying the actions of their elders.

Back at their rock shelter base the family watched the onset of the monsoon, its water causing a great surge of life. Taking spare bamboo

Storm over East Alligator.

shafts and axe blank some of the family walked up the valley passing many sacred places and on the way crossed the dreaming path of two creator beings who had taken an important ceremony from the coast to the stone-country people. These two beings had 'put' images on the walls of several shelters including their own home shelter. The two fading paintings of these male and female wallaroo beings danced on the camp-fire lit wall and constantly reminded the family of their past and present.

The monsoon season, *Gudjewg*, had begun when the family next visited the wetlands, now a vast inland sea of floodwater, to collect goose eggs. With the wetland people, they feasted for a fortnight and then, taking some eggs for possible trade they made their way back. No more than half the eggs were taken from any nest. With half a million or more geese in the swamps, the impact was not very great.

Return was leisurely. Carrying bamboo and eggs they headed for the great imposing red bluff of Nourlangie where several days would be spent in the cathedral-like *Anbangbang* shelter made when the front of the mountain had rearranged itself. On a wall nearby one of the great grandsons later painted the famous Nourlangie frieze. The family also joined with the local people and hunted flying foxes in the paperbark and pandanus forest fringing the billabong.

Moving around the escarpment of Mt. Brockman the travellers paused near *Guri Birang Doi*, to retell the story of an accident that the blue-tongued lizard had in the creation time. If the creek were in flood while they were crossing the plain to Deaf Adder, they would call on the Rainbow Snake to send the dogs to help them across. This continual reliance and reference to the ancestral dreamtime built and continuously reinforced the traditions of the Aboriginal family.

Spear grass now stood above their heads and was beginning to turn yellow when they once again visited the stone country to share and trade the bamboo. While there, the violent downpours of the last storms of the wet broke down the tall fragile spear grass stems . . . the season (April) of the knock-em-down storms, *Bang-gereng* had arrived. On this trip the men attended to ceremonies relating to the successful propagation of certain plants and animals at special dreaming sites.

By the time the party had returned the high-pitched skirl of the little green grasshopper, *Yamidj*, was signalling the end of the storms and the start of the season of plenty, *Yegge*. *Yamidj* 'told' the people that the 'cheeky yams' were ready. These yams were poisonous and needed special treatment before they were suitable for eating. All around the bush was alight with wattle and the wooly-butt trees were bursting into their orange flowers, which during the day attracted hordes of screeching lorrikeets, and at night, squabbling flying foxes. Honey from many native bees' nests filled the paperbark store 'pots'.

The dry season was with them now and at night they camped on the sandy banks of *Djuwarr* lagoon made by the Rainbow Snake when it split the nearby cliff and moved up the gorge to sleep in the deep pool below the falls. *Wurrgeng*, the cold weather season, is heralded by sparkling dewy mornings and the camp begins to talk about the year's big walk which would begin by following the path of *Yamidj* towards Jim Jim and Twin Falls. In the dreamtime the green grasshopper followed this route, his digging stick leaving behind rings of stones where he planted the water yams. Along this unstable and towering cliffline lived a harmful spirit, *Malindji* which closed caves on people who were thoughtless enough to sleep in them.

Where the travelling was difficult due to the thick undergrowth, the firestick was put to the grass and the cool fire cleaned the country, going out as the dew formed. At the sight of a smoke column a wheeling pyramid of black kites dropped silently from time to time into the smoke and flames to take fleeing insects, reptiles and small mammals. They

Right After the 'knock-em-down' storms.

Kakadu — man and landscape

passed the goanna dreaming gorge, the red ochre dreaming and the place where one of the 'first people' made hooked spears, and, as he hurled, would call out the languages the people would speak, where the spears should fall. Towards Barramundi Creek they passed through a great grassy area made by the emu during creation and here they successfully hunted the emu. Nearby the ground yielded up dolerite axe blanks and a large piece of haematite which would be used later to paint new designs on rock shelter walls.

When they reached the upper Katherine River they found a gathering of peoples from a wide region preparing to put together song cycles, stories and ceremonies of the creation period. This was a very special place with stringent rules as to how the people should conduct themselves. Like mile pegs the route from the ceremonial site passed bees' nest, white cockatoo, white apple, green plum, sugar glider and fly's head dreaming places. They were now back across the plateau and east of Deaf Adder on the headwaters of the East Alligator River, which cut deeply into the joints of the sandstone and ran into deep wild gorges with vertical walls towering over rapids and deep pools. Freshwater crocodiles, black bream and the saratoga were eaten along the way. Down river they walked and swam until the gorge opened out into a broad valley. In a cave containing paintings the men lit special smoky fires to ensure a plentiful future supply of yams.

Near Oenpelli the Bardmardi crossed the river for the last time by the same crossing used by the creator of all people in this region on her journey of creation.

Now on the return loop they camped by the Djabiluku waterhole feasting on barramundi, catfish and agile wallaby. There were places near here known to be dangerous ... where to camp would mean sickness from sores and bleeding. At the end of the Djabiluku outlier they came again upon another campsite of the mosquito ancestor and followed its track to Mt. Brockman where they speared a black wallaroo as it came down the slope to drink at the spring.

The days were warming up now and the walking was frequently on burned ground as the party crossed the valley from Brockman to a most sacred place, the Lightning Dreaming. The children knew the rules of behaviour here for they had visited it on numerous occasions and by experience knew of the strength of the lightning spirit's activity in the storm season.

Gurrung, the tiresome hot-dry season, had begun by the time they had returned to the cool shade of the big weeping paperbarks overhanging Deaf Adder Creek. September and October were wiled away fishing and hunting the fat, long-necked turtles. It was now time for the neighbouring clans to visit the Bardmardi for ceremonies in the Deaf Adder Valley.

One could write much more about the depth of knowledge the Aborigines have of the natural community to which they belong but it should be quite apparent now that they are continuously learning the processes of living and that the landscape itself is not only teacher but text book as well. It is little wonder then that when Europeans arrived they had no capacity to understand what they described at best as simple people. They looked not for a harmonious relationship but for evidence of ownership, of manipulation, of control, not a relationship more akin to membership ... many still do.

How is it possible to cut up such an integrated system into artificial units without totally destroying the unity which is its reality? Is it any wonder that an Arnhemland Aborigine upon seeing the parcelled up landscape around Canberra suggested sadly that 'this land has lost its music'?

The Europeans

These photographs were taken by an English cyclist, Ryko, and are from the Jack Stokes collection, Australian Archives.

In a social and economical point of view, it is difficult, if not impossible, to over-estimate the importance of the discovery recently made of an all but boundless extent of fertile country, extending to the north, soon to be covered with countless flocks and herds, and calculated to become the abode of civilized man. In its political aspect, the possession of an immense territory, now for the first time discovered to be replete with all those gifts of nature which are necessary for the establishment and growth of a civilized community, cannot be regarded as a fact of small importance.

(Testimonial speech by the Hon. Speaker of the Legislative Council, N.S.W., Dr. C. Nicholson, in *Journal of an Overland Expedition in Australia* — L. Leichhardt, 1847.)

Such was the response of 'civilized man' to Leichhardt's journey of discovery, which terminated in the first traverse of what is now Kakadu National Park. Here is the classic nineteenth century attitude of the colonizing Briton, the ultimate recipient of the gifts of nature and God's gift to the 'uncivilized hostile native,' under whose stewardship these gifts of nature had been maintained for perhaps 50,000 or more years — 2,000 generations.

Kakadu National Park is more than a fascinating landscape, displaying a richly varied population of wildlife, along with probably the largest number of Aboriginal art galleries in the Northern Territory, it is the home of the traditional owners. Their story gives us insight into the processes involved in the lines — 'replete with all those gifts of nature which are necessary for the establishment and growth of a civilized community.'

However, radical change was on the way even before Leichhardt's journey. After crossing the East Alligator River, he 'encamped at a good sized waterhole at the bed of this creek, the water of which was covered with a green scum ... the dung and tracks of the buffaloes were fresh.' Just the day before (10.12.1845), Aborigines smelt the explorer's meat and said in disgust, 'You no bread, no flour, no rice, no backie — you no good!'

No one is quite sure when the Buginese fishermen began to ride the north-west monsoon in their praus to the northern Australian coast. These first 'boat people' gathered in their thirty tonne, high bamboo and wooden vessels at Macassar, and took on rice, water and firewood for the 'downhill' voyage. Once on the coast, they would fish the shallows for trepang from dugout canoes, and return to camp, set up on beaches and behind headlands with rows of cooking pots. Aborigines tell and sing stories of these people who left behind some of their words, and the concept of the dugout canoe. There was little or no competition where the cultures met on the frontal dune and in the shallows. Aborigines didn't use trepang, and the Bugis didn't need assistance and they didn't try to implant their religion. With the fishing complete, a plate of food was lowered to the sea floor for the purpose of 'rising the wind' and the ceremony was completed with song and dance. True to form, the south-east winds swept the loaded praus back to Macassar — only broken pots, fire lines, wells, ritual posts and corpses of their less-fortunate shipmates were left. As if to mark their passing, carelessly dropped seeds from the Tamarind (*Tamarind indicus*) pods brought in as a staple food, germinated and grew into deep green living monuments. (MacKnight, 1976)

So far, the impact of the outsiders meant an enlarged vocabulary, an

enriched story and art resource for the Aborigines, and a new tree which provided dense shade and pods to eat. However, the indirect results of the Macassar visits were to have a profound effect.

Matthew Flinders, followed by Phillip Parker King, who made a systematic survey of the Arnhem Land coast from 1818–1822 and who named the Alligator Rivers and Barron and Field Islands, encountered fleets of praus. Flinders recorded sixty and King was told in Kupang that 200 praus visited the coast annually. More than a thousand men fished and preserved trepang.

Trading buccaneer William Barns put to the British Government a year later, the idea of opening a port in northern Australia to act as another Singapore to absorb the trade of the 'Spice Islands', and to use the Macassan fishermen as the tool of trade to outplay the Dutch — Singapore at one end, and an Australian port at the other. (Blainey, 1966)

The first attempt at Fort Dundas, Melville Island, opened in 1824, but by 1829 had not attracted one Indonesian visitor. Fort Wellington at Raffles Bay, Coburg Peninsula, was next tried and officially lasted less than a year, but the command to abandon did not arrive until 22 months had passed, and by then the praus were arriving in port in greater numbers. But close it they did, and Port Essington was opened. Poor communication, disease, scurvy, heat, and unco-operative natives kept the settlements in a state of flux. However, one animal was settling in.

The water buffalo — 'the cow crossed with a bulldozer' — was the first successful settler, arriving from Timor in the years 1824 to 1829 to supply the settlements with meat, and offering motive power adapted to tropics and mud. With each settlement failure, the buffaloes were released to fend for themselves. Along with the buffaloes was released a less aggressive animal — the Banteng* cattle, which have reached herds of considerable size (1,000+ animals) on Coburg Peninsula. It was apparently unable to run the crocodile blockade in the swamps, but the buffalo, agent of the unsuccessful settlers, and larger, heavier and better adapted to wetlands, made the break into the Northern Territory proper. Leichhardt, in 1845, travelled through the first herds to do this (north of Oenpelli).

The buffaloes took up home ranges, each of which included a number of points fixed to buffalo behaviour — toilet area, drinking point, feeding area, a wallow, rubbing trees, and a camping area. In the wet season the animals had much water to share. As the 'dry' intensified, this resource shrank, and more and more of the buffaloes converged on this diminishing area. It is this water less than a metre deep, which is so productive of sedge meadows and aquatic macrophyte stands in the Kakadu wetlands, and which is the base of major native animal populations, and upon which the Aborigines depended for much of their food — food in such quantity and quality that it was able to support the gathering of many clans to reap the bonanza, and to conduct important ceremonies.

Typical buffalo behaviour in the receding waters of the dry season severely degraded the wetlands. As the waters receded into disconnected ponds, the animals linked these with wading channels, and then these to the rivers or coast, breaking and channelling the levees. The backswamps, which once held fresh waters for long periods began to drain quickly, and in many places salt water actually backed into and killed many hundreds of hectares of *melaleucas* and freshwater swamp. With lower water levels, wallows were selected under the shade of trees along the banks of billabongs and rivers. Banks broke down, and trees tumbled into the waters to be swept away in the succeeding wet seasons. As if to exhibit their utter contempt for the natural system . . . 'buffaloes will often retain their faeces and urine until they are immersed in water.'

Left Twin Falls in the dry season.

*These herds of Banteng are the largest herds of this species still living in the wild — a species listed in IUCN's Red Book of threatened animals.

(Cockrill, 1974 p. 279). Then the lot is thoroughly mixed into a turbid solution in the succeeding wallowing. Dissolved solids are increased by buffaloes probably as much as five times, suspended solids forty times, calcium six times, magnesium three times, potassium forty times, and phosphate thirty to forty times, to mention a few changes. The result is a heavily eutrophic pond, or as Leichhardt wrote, a pond 'covered with green scum.' By October, the floodplains lie drained, cracked, pockmarked with wallow craters, and the vegetation gone.

An astute Captain Carrington, making a resource evaluation for the South Australian Government of the lands between the South and East Alligator Rivers in 1886, wrote:

Half a mile below the ship's anchorage the jungle was within half a mile from the river and was fringed with Tea-tree swamp, with a little water remaining; here large herds of Buffalo numbering several hundreds were seen daily.

. . . these animals are increasing rapidly and if some preventative measures are not taken it is only a question of time when they will so increase as to become a serious evil.

In the meantime, some buffaloes were sheltering and developing rubbing stations in art galleries, rasping off the art, and plastering it with mud.

Buffaloes have a social impact as well as these very obvious physical problems. Betty Meehan of A.N.U.'s Research School of Pacific Studies (1980), working with a clan of coastal Aborigines of the Blyth River, makes the comment —

*More recently, buffaloes have begun to move out onto the black soil plains of Anbarra land and in 1978 and 1979 herds of up to 30 animals were observed grazing on the plains. These animals have already fouled many of the inland waterholes and occasionally they pollute even those associated with coastal home bases. They also trample wetland edges and batter the small jungle patches reducing the latter in many instances to a single banyan tree (*Ficus*) or* Corypha *palm devoid of the shrubs and vine thickets containing yams and other plant foods which make up the normal structure of these floral associations in untrampled areas. Their presence in the region and their effect on the environment has already caused major changes in Anbarra hunting strategies. People, including male hunters, are still afraid of these beasts and are only gradually acquiring the skill to hunt them successfully with guns. If buffalo numbers continue to increase at the present rate and features of the landscape continue to be altered and in some cases destroyed by them, then within a decade or two, perhaps even less, the focus of Anbarra subsistence may have changed dramatically from one in which the emphasis was on ecclecticism to one in which they focus onto one species. At the moment most members of the Anbarra community (men, women and children) contribute some food towards the total diet. If buffalo were the mainstay of their diet the community would then become dependent on a few young hunters.*

One of the great strengths of the coastal economies to date has been their diversity. If the buffalo population continues to increase without check then much of the basis of their diversity will be under threat and though, in the short term the promise of mountains of red meat may appeal to the Aboriginal population, in the long run its acceptance will mean a deterioration in the quality of their life.

It would seem to me that this kind of dependance on the buffalo began many years ago in the Kakadu area, once heavy rifles became available with the advent of the buffalo skin industry. There were certainly animals available at the time of Captain Carrington's report (1885–6). In 1979 I saw the Cannon Hill outstation people waiting patiently on the rocks by the lagoon for the twice a week delivery of buffalo meat by whoever had taken the animals that week.

The early phase of buffalo shooting — a spine shot immobilised the animal keeping the skin warm until the skinners arrived. The hunters were expected to feed off the catch. (Photos: Jack Stokes collection, Australian Archives)

While the buffalo, which has recently been referred to as 'the Devil's disciple' was permeating the physical and social environments of the Kakadu region, other changes were taking place which were to have greater social than physical impact.

Following the collapse of the first four settlements (Fort Dundas, Fort Wellington, Port Essington, and Escape Cliffs), it was Goyder, the Surveyor General of South Australia who ultimately selected the site of Darwin (first called Palmerston) as the service centre for the north. Speculators and hopeful farmers and pastoralists took up blocks but soon found the area unproductive.

By 1872, Australia was linked by cable via Darwin to the world. This settlement was permanent. Apart from Government and British/Australian Telegraph Company employment, Darwin was made up of sailors, pearlers, fossickers and traders. To paraphrase the Port Essington Aborigines, 'plenty bread, plenty flour, plenty rice, plenty backie, Darwin very good.' So the Aborigines drifted in, some to be trained and paid up to 35/- per week, most as town rouseabouts receiving 2/- per week together with keep of a kind. It was not unusual for Aborigines to arrive in the morning offering to work, carry wood and water and the like, and then just as suddenly go bush. However, there were other attractions in Cavenagh Street, as the *Northern Territory Times* of 1.5.1889 recorded:

Almost every coin that is given to the blacks finds its way to the opium dens, and even those who do not smoke are forced to hand their earnings over to their opium-soddened relations. It was only last Saturday that, immediately after giving my boy a shilling, he was pounced upon by a lubra, who said that if she did not get a smoke she would die.

There were, of course, many hundreds of Chinese in Darwin, most being in Australia to make a fortune in the goldfields, but drifting always north towards Darwin and the lately discovered fields of the 'seventies and 'eighties out from the north-south railway. The construction camps for the railway and the diggings also lured the Aborigines —

No doubt you came across those settlements on the railway line. You found about three white women to every hundred white men, and you must have noticed how those white women were treated. Why, man, they're venerated as saints. (Warburton, 1944)

Aboriginal women could be bought for 2/- or a piece of tobacco (Warburton, 1944) or less by Europeans but there were stringent regulations concerning co-habitation with Chinese, and which, of course, were broken.

From about 1870 to 1910 most minerals were found and worked by the industrious Chinese and at the same time gangs of Chinese timber cutters were cleaning almost every stand of cypress pine from the Alligator

Rivers region. Even Barron Island at the mouth of the South Alligator had been completely logged by 1885. (Carrington, 1885–6). Finally, the Chinese were banned from working new mining fields in 1915.

In this spasmodic splurge of mining activity towns boomed overnight, land was engulfed, the traditional people were pushed aside, sacred sites violated, natural food resources wrecked or dispersed. Local Aborigines were bewildered and demoralised. Only in a few areas were the men allowed to labour in the mines, though women were welcome as 'domestic' helpers. Moline was one place where Aborigines washed and crushed the ore and on the weekends went with the whole camp to Jim Jim to get fish, ducks and geese and in season, eggs. Aborigines too, when asked the right question were highly adept at locating ore bodies.

Under its deep mantle of silt the wetlands of Kakadu were temporarily at least protected from the miner. But these wetlands supported large numbers of buffalo and it was Paddy Cahill from Oenpelli who commenced the hunting of buffalo for hides from horseback in the 1890s. Fred Smith, who became established in the abandoned Kapalga mission station, worked the plains of the South and West Alligator Rivers. This was an activity better understood by the Aborigines, although by now demoralisation and disease had greatly reduced the population. By 1900, the buffalo resource near to Oenpelli had almost been shot out.

Ultimately, shooting teams comprised foot shooters and mounted shooters — the infantry and the cavalry. Mounted shooters worked ahead of the team and paralysed animals with a spinal shot which kept the skin warm and soft so that when the skinners arrived, it could be removed. Camps were established near billabongs so that hides could be washed, some camps at least taking over traditional dry season camps of the clans. Not only did the Aborigines fit these enterprises more easily because they were hunting activities, the standard of living within the camp was common to Europeans and Aborigines — poverty was endemic. Diet was mainly damper, buffalo meat, treacle, jam, tea and 'bush tucker' while accommodation was a bough or paperbark shed.

At the end of the shooting season, the Aborigines received a wet season supply of flour, sugar, tea and tobacco. Usually this was consumed quickly at a get together just as those other itinerants, the shearers, frequently 'blew their wages' at the first town. Then it was a spell of traditional living with visits to the stations for supplies of plug tobacco, the universal lure.

The outcome of all these activities and others, usually as debilitating or worse, was a dispossessed people with lost self-esteem, many suffering disease, with shattered morale as they saw their long-held traditions and sacred places defiled and ridiculed, looked on that great escape mixture, alcohol and tobacco. The stage was set for . . . the saviours.

The Northern Territory Native Industrial Mission was first to begin in 1900 when it established buildings, eight donkeys, twenty goats, gardens of bananas, coconuts and pineapples at Kapalga. They found this site too open to outside influences and in order to isolate the children from adults 'contaminated' by Chinese miners, the missionaries decided to move to Greenhill Island in Van Diemen Gulf just south of abandoned Port Essington. Its fate is not clear, but perhaps the message struck home to Aborigines that here was yet another pressure aiming to break up the remnants of the clans, and so they made themselves 'unavailable' for saving.

A whole generation went by before the Church Missionary Society accepted the offer of Oenpelli from the Administration. In September 1925, the church was once again established in the area. It seemed clear to the missionaries that Aboriginal society had to be restructured and this would involve a rebuilding of personality and lifestyle.

Early on, in the days of Paddy Cahill and the Agricultural Research

Top Quartz reefs on Koongarra Fault, Mount Brockman.

Bottom Rock paintings, Nourlangie Rock.

Station, many Aborigines from perhaps ten or so clans visited Oenpelli to socialize, to work, and to buy tobacco. So the site was potentially a place where the missionaries could sift through a significant number of Aborigines for potential converts. With great sincerity, Alf Dyer the first missionary, persuaded mothers to bring in their children for education.

The Church Missionary Society's method was to attempt to create a community of independent, small farming, nuclear families. This involved the conversion of a lifestyle held for perhaps a thousand or more generations. The system appeared to work well in the early years, the people seemed to be 'busily' occupied working with a purpose. However, to achieve this, the missionaries not only controlled the direction of labour, but also programmed the hours of the Aborigines between church, school, dormitory and garden. The people could desert Oenpelli if they so desired, but it did provide food and a kind of insulation from the pressures of a world which had become less and less a place where natural order existed and more and more intervention by the unpredictable European, his feral animals and disease.

However, it was difficult for the Mission to compete with the financial rewards being offered by the revitalised buffalo industry. Returns from work in the industry meant access to alcohol and tobacco, if the Aborigine wanted these things. Dyer had been realist enough to give the residents of Oenpelli tobacco as part payment of wages but this was stopped and tobacco banned by his successors. Cole (1980), makes the comment 'The "no tobacco" rule left many men and women unstable.' This action promoted the use of the 'Border Store' on the East Alligator River west bank and the Cannon Hill buffalo camp, where tobacco and alcohol were available. Quite a number of the people who tried mission life were driven from it when the children were taken from them and placed in dormitories. These and others camped over the hill, interacted with the resident Aborigines obtaining permission to stay, but not becoming involved in the mission programme.

Cole (1975) comments that after visits to the Border Store —

The resultant misbehaviour, fighting and absenteeism began to threaten the very existence of the Oenpelli people. At the same time, paradoxically, this decade witnessed an astonishing number of baptisms . . .

The Chief Protector of Aboriginals, Dr. Cook, attacked the missions because they 'neglect the detribalised Aboriginal, devastated by disease, disrupted by the impact of white civilisation and lost in the new social order' and instead select 'the native, living happily in his own country, for the most part secure from molestation and the impact of white civilisation' for their own brand of deliberate social disruption. He was not popular in mission circles!

Cook stressed the importance of large reserves as a sanctuary for Aborigines. His attitude was at least Aboriginal-oriented. Early reserves were established primarily as places to push Aborigines into to keep them out of town. (*N.T. Times* 14.12.1900)

As early as 1892 a number of small reserves were gazetted west of the Alligator Rivers. In 1931 the most important Aboriginal reserve of all was proclaimed — the Arnhem Land Aboriginal Reserve incorporating the reserve around the Oenpelli Mission. A notice was gazetted in 1936 providing for arrest without warrant for any unauthorised persons entering an Aboriginal reserve. In the same year the smaller reserves were reviewed, the original Woolwonga Reserve was rescinded and the name given to a new area east of the South Alligator River known to be important to Aborigines as a major food and ceremonial place. The area could also be identified.

On the recommendation of Professor Baldwin Spencer, a compound was erected at Kahlin Beach for those people employed in Darwin — a curfew was in force. In an attempt to set social boundaries on the interaction between blacks and whites and between mixed races and both, complex rules, economic disincentives and social restrictions further seriously ruptured relations between the people. These included: no loans from the Administration if 'coloured labour' was employed (1935); absolute prohibition against the employment of Aborigines by Chinese; Europeans had to be licensed to employ Aborigines; three to six months in gaol for single Europeans employing Aboriginal women; a single European could be charged with co-habitation for speaking to an Aboriginal woman in the street. These naive rules were fraught with opportunity for bent and selective operation.

While this social manipulation by the Administration was having effect in the towns, the mining fields, and along the railway, the Alligator River Region was protected from it because police patrols usually reached Jim Jim only once or twice a year. However, these and other attacks on human dignity would fuel the bitterness of the attitudes widely and publicly expressed concerning the issue of land-rights and the management (from policy creation to execution) of the lands known as Kakadu National Park. One could very well take a leaf from novelist James Michener and call this whole chapter *From the Farm of Bitterness*.

The coastal Aborigines had been working with Macassans and Japanese pearlers for many years before the Japanese declared war. To prevent possible aid being given to the enemy, people were moved away and placed in settlements at Adelaide River, Katherine, and in 1943, Cullen River near Pine Creek. Those people in the buffalo camps were allowed to stay on because of the need to maximise food supplies. Other Aborigines played hide and seek with the army for years. But in the few years that the people from Pine Creek, Daly River, Mary and Alligator Rivers were kept together with little food, they learned to trust each other and in the future they were to learn that together they had much more power in dealings with Europeans.

With the end of the war, Aborigines and Europeans associated with the Alligator Rivers returned, and hunted buffalo and crocodiles with renewed vigour, but too soon the hide market crashed as leather generally gave way to synthetics, and crocodile populations were decimated. Echoes of conservation arguments began to be heard up north. Uranium was discovered and mined at numerous deposits about the upper South Alligator River valley and exploded in the British atom bomb experiments. In the meantime, the American-Australian Arnhem Land Expedition results were being communicated. Northern Government infrastructure was turning over and some at least of those public servants had been impressed. Exotic stories of barramundi, crocodiles, birds, billabongs and fabulous art sites were trickling south via tourists. A certain pride in the Top End was developing and the twentieth century was catching up, as was the idea of self-government.

By 1960, however, there were very few Aborigines left in the Alligator Rivers region. A brief respite from unemployment occurred when buffalo meat became a profitable crop for pet food and for human consumption. Mudginberry attracted a temporary Aboriginal population of about 130, until award wages were granted in the late 'sixties. Improved wages meant fewer positions and those people who remained in camp by the billabong became dependent upon the few left with work. The luckless ones drifted to the Border Store on the East Alligator and drowned their woes in alcohol paid for by social welfare payments.

Events in the south-eastern part of the continent were creating a public awareness of the plight of Aborigines and certain highly motivated and effective Aboriginal spokesmen were at work establishing credibility for the Land Rights movement. The overwhelming support given by the public of Australia to the referendum on Aboriginal issues greatly reinforced these moves. The Aboriginal people were again re-stating their

Right Horns and dried salted skins at the loading wharf on Adelaide River. (Photo: Jack Stokes collection, Australian Archives)

kinship with the land and all things natural.

So in 1970, if ever an Alligator Rivers Aborigine looked back over the past 120 years, over just four generations of European occupation, of feverish activity, and sought to find what the achievements were of the period, he would see

- an environment considerably broken by feral animals
- his people decimated by disease — 2,000 persons reduced to 80 in the Alligator Rivers region, by smallpox, leprosy, V.D., T.B., malaria, flu, measles
- numbers of bankrupted industries
- a dispossessed people resorting to escape in alcohol.

In fact, there was almost nothing positive in the region to show for all that destructive European activity.

Typical of many, one young Aborigine about to be lifted from a hunter/gatherer economy to a central and effective political position arguing his cause with international business, lawyers, the Australian Press Gallery, and the Prime Minister, was then coming to grips with his traditional beliefs —

When I was young man I was accepted to be initiated, enabling me to tell and sing and to perform any ceremony.

In all these ceremonies, I learnt things that only tell about the tribal land, how it is talked about, sung, dance, paint and most of all feel as though the land is another you.

One day I went fishing with dad, as I was walking along behind him I was dragging my spear on the beach which was leaving a long line behind me. He told me to stop doing that. He continued telling me that if I make a mark or dig for no reason at all I've been hurting the bones of the traditional people of that land. We must only dig and make marks on the ground when we perform or gather food.

The land is my backbone. I only stand straight, happy, proud and not ashamed about my colour because I still have land. The land is the art. I can paint, dance, create and sing as my ancestors did before me. My people recorded these things about our land this way, so that I and all others like me may do the same.

I think of land as the history of my nation. It tells us how we came into being and in what system we must live. My great ancestors who lived in the times of history planned everything that we practise now. The law of history says that we must not take land, fight over land, steal land, give land and so on. My land is mine only because I came in spirit from that land, and so did my ancestors of the same land. We may have come in dreams to the living member of the family, to notify them that the spirit has come from that part of our land and that he will be conceiving in this particular mother.

My land is my foundation. I stand, live and perform as long as I have something firm and hard to stand on. If there is a flood on my land I will have to swim and all Gumatj clan will have to swim, but not for long, we will surely perish, then we will be just like thousands of other people whose lands have been stolen away from them.

We will be the lowest people in the world, because you have broken down my backbone, took away my arts, history and the foundation. You have left me with nothing.

Without land I am nothing. Only a black feller who doesn't care about anything in the world. My people don't want to be like you!
(James Galarrwuy Yunupingu.)

The stage is now set to begin to understand the intricate, sometimes cruel, sometimes treacherous play of politics, economics, propaganda, and social pressures which finally gave the Australian nation a rather unusual National Park, and the world a major heritage area. It is not altogether a story to be proud of. (*Summer 1982, Winter 1983*)

Movables of heritage significance

Rosemary Hassall
Photographs by D. Seeto

Rosemary Hassall is a professional Appraiser. A graduate with majors in Fine Arts and History, an Approved Valuer (Commonwealth), she is co-author of a textbook Things of beauty, rarity and distinction, *Thomas Rowland, 1984.*

The approach of the 1988 Bi-Centennial year has led to an upsurge of interest in the Australian national heritage. Whilst in recent times, protection of the natural heritage and of fixtures such as buildings has captured most attention, there remains a need to protect and conserve movable property constituting part of the national estate. This article examines some aspects of law and practice relating to works of art, historic relics and the like which prom-

Australia in Trust

ise to become of increasing importance in Australia. These include measures to stem the illicit export of items of significance to the national heritage together with other measures tending, directly or indirectly, to the protection and conservation of such items in the interest of the Australian public, both now and in the future.

The existing Australian law and practice in this regard is inadequate. The *Customs (Prohibited Exports) Regulations* made under the Commonwealth *Customs Act* 1901 prohibit the export, without Ministerial consent, of limited categories of items including certain native fauna, documents recording transactions between Australian Aborigines and early settlers and explorers, Aboriginal artefacts, certain coins and tokens made prior to 1901, contemporaneous records connected with, and goods owned and used by, persons associated with the discovery and early settlement and exploration of Australia.[1] It is not simply a matter of these *ad hoc* categories being inadequate: no guidelines for the exercise of the Ministerial discretion have been laid down. Moreover, the legislation is not specifically directed to the protection of the heritage, and so it is not surprising that it does not establish any machinery or system to ensure proper control.

In 1975, the Pigott Committee of Inquiry recommended that the Commonwealth should introduce specific legislation to control the export of items of major importance to the national heritage of Australia. More recently, that view was echoed by Mr. Justice J. S. Lockhart, a Member of the Board of the International Cultural Corporation of Australia.[2] The present writer believes that Australia would do well to legislate along the lines of Canada's *Cultural Property Export and Import Act* 1975, which contains specific provisions for the regulation of the export of items of national heritage value and the import of foreign cultural property. The Canadian Act sets out appropriate procedures for review of decisions and compensation for owners and *bona fide* purchasers.

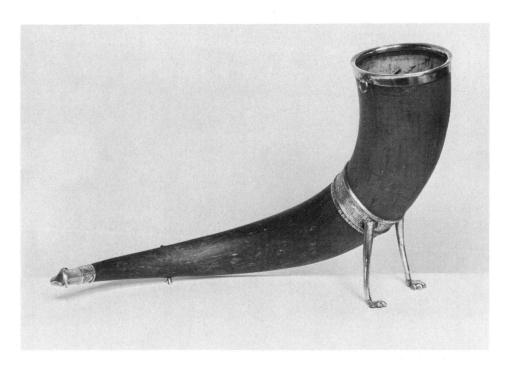

Above **The Pusey Horn, Victoria and Albert Museum.** (Crown Copyright)

Right **'Collectables'.** (Photo: D. Seeto)

The Canadian Act implements the UNESCO *Convention on the Means of Prohibiting and Preventing the Illicit Import, Export and Transfer of Ownership of Cultural Property* 1970, which Australia has not, at the time of writing, yet ratified.[3] However, an effective Australian statute similar to that of Canada establishing a National Register or Export Control List in respect of movables of heritage value need not necessarily await the ratification of the Convention by Australia. At least insofar as export and import is concerned, the Commonwealth could rely upon its clear and undoubted powers, such as the Customs power, to support such legislation without resort to the external affairs power.[3A] The Canadian Act establishes procedures for statutory accreditation of appraisers with respect to wide categories of property of heritage significance, including works of art, historic relics and the like. Provision is also made for review of administrative decisions made under the Act, to ensure fairness and justice.

In the period since the passage of the *Museum of Australia Act* in 1980, attention has been focused on instances where things of importance to Australia's history and heritage have been lost to the nation. There has been particular concern at the threat of loss of items connected with explorers, for example, the removal of Mawson's gear from Antarctica. International attention has also been drawn to the protection of movables of heritage significance by the efforts of Melina Mercouri, the Greek Minister for Culture and Sciences, to have returned to Greece the marbles removed from the Parthenon by Lord Elgin.

Very recently, we saw the manifestation of wide public interest and concern that an oil portrait of Captain Cook by an English artist be purchased by an Australian public institution: in the event, public relief was expressed when it was revealed that the successful bidder for the painting at auction was an Australian collector. It is to be hoped that such concern will be reflected in specific national heritage export control legis-

lation sooner rather than later, and certainly before 1988.

On the whole, Australia lacks specific legislation directed to the protection and conservation of movables of significance to the national heritage. However, such items receive some degree of protection under legislation for the protection of the heritage generally.

At Federal level, the relevant legislation is the *Australian Heritage Commission Act* 1975. Under this Act the Australian Heritage Commission has functions which include advising the Commonwealth on matters such as the conservation, improvement and presentation of the 'National Estate', a Register of which is established by the Act. The Act defines the 'National Estate' as consisting of places that are part of the natural or cultural environment of Australia having 'aesthetic, historic, scientific or social significance or other special value'. Further, under the Act, 'places' is given an extended meaning which includes buildings or other structures and, importantly for present purposes, 'equipment, furniture, fittings and articles' associated or connected therewith. Another piece of Federal legislation, the *Historic Shipwrecks Act* 1976, extends protection to certain shipwrecks and relics of historic significance.

At State level, examples of the extension of heritage legislation to movables are to be found in the Victorian *Historic Buildings Act* 1974 and the New South Wales *Heritage Act* 1977. Under the former, 'building' is defined to include a 'building, work or object or any part thereof or appurtenance thereto' whilst under the latter, 'relic' is defined to include deposits, objects or material evidence relating to the early European settlement of New South Wales. Also noteworthy among State legislation are the Queensland *Aboriginal Relics Preservation Act* 1967 and the Northern Territory *Native and Historical Objects and Areas Preservation Act.*

Where hidden gold or silver coin or plate is uncovered, it is a moot point as to how far the ancient law relating to the Royal Prerogative in 'Treasure Trove' still applies in the various Australian jurisdictions. Traditionally, an inquest into a find of Treasure Trove was conducted by the Coroner. In Australia today, Coroners operate pursuant to statute. In the United Kingdom, the recent popularity of electronic metal detectors with weekend treasure-hunters has led to some remarkable finds, some difficult cases for the Courts,[4] and, very recently, a significant statutory extension of the Treasure Trove concept to encompass the wider category of 'antiquities'.[5] Whilst the comparatively short history of European settlement in Australia and the vastness of the continent mean that cases of Treasure Trove or finding of other precious objects are fairly uncommon, this is not to say that the principles relating to these matters are insignificant for us.[6]

It is submitted that there is a case for the establishment of a National Register of Movables of Heritage Significance. This could be done in conjunction with or as an integral part of, specific legislation to control illicit trafficking in historic relics and the like. The existing Register of the National Estate goes some distance in this direction. However, as noticed above, a connection or association with a particular 'place' is necessary and many items of significance to the national heritage may have no such relation to any particular place.

An area of law that is of special significance in relation to movables of heritage value is the jurisdiction of Courts of Equity to order the specific delivery or return of chattels instead of merely awarding damages to the dispossessed owner. This jurisdiction was confirmed in the celebrated case of *Pusey* v. *Pusey* in England in 1684.[7] There, the Court ordered that the Pusey Horn, an heirloom consisting of a medieval drinking horn mounted with silver and said to have been given by King Canute in granting title to the manor of Pusey, be delivered to its rightful owner. The horn is now in the Victoria and Albert Museum in London.

In other English cases, the principle was applied to certain jewels, unusual articles of plate, and paintings. The principle was received into Australia at British Settlement in 1788, along with the bulk of the English Common Law, and the Supreme Courts in Australia today possess the same equitable jurisdiction, as well as a like discretion to order specific delivery under Sale of Goods legislation.

Because of the early history of the principle, it has sometimes been thought that its application is confined to articles of 'unusual beauty, rarity and distinction'.[8] However, whilst it is true that under both the equitable and the statutory jurisdictions, Courts will not order the specific delivery of ordinary articles of commerce of no special value or interest, the principle can be applied, and has been applied, to some very mundane articles indeed. The crucial requirement is that the article have some special value or interest in respect of which mere damages would not be adequate compensation.

These remedies are quite commonly invoked in the Australian Courts and they can assume particular importance where unique or exception-

Example of a brass King Plate, which was presented to Aborigines in the 19th century.

ally unusual and valuable articles are involved. For instance, the equitable remedy may have useful application to Aboriginal artefacts or other historic relics.

The Taxation Incentives for the Arts Scheme, first introduced by the Federal Government in 1978 on a trial basis and confirmed in 1982 as a permanent feature of the *Income Tax Assessment Act* 1936, represents a development in Australian law of major importance to the national heritage, particularly as it relates to movables. Essentially, the Scheme allows taxpayers to obtain a taxation deduction for donations of works of art or other movable property to The Australiana Fund, Artbank, public libraries, public museums and public galleries in Australia.

The importance and great merit of the Scheme lies in the fact that it encourages private individuals to donate works of art and other articles to the permanent public collections in Australia, thereby enabling the Australian public to have access to works which might otherwise have remained in the private domain and seen by only the few. Difficulties encountered during the trial period of the Scheme, such as attempts by some to use the Scheme improperly for tax avoidance purposes, have been overcome by amendments to the legislation made in 1982, one of which formalised the procedures for the accreditation of duly qualified persons as Approved Valuers for the purposes of the Scheme. As a result, the Scheme has extra significance in that it provides a statutory form of accreditation, operating throughout Australia, of duly qualified valuers or appraisers of chattels such as works of art and the like.

The Scheme is based on a precedent under United States Internal Revenue legislation. Similar schemes have operated in relation to death and estate duties in the United Kingdom. No direct cost to the public purse is involved: an important consideration in times of economic constraint generally and in the arts field in particular. It should be noted that the Australian Scheme benefits not only public institutions such as art galleries but also The Australiana Fund, which provides Australian pieces to furnish official residences under the control of the Commonwealth, and Artbank, a company set up by the Commonwealth to encourage Australian artists by acquiring works for exhibition in public buildings under a hiring scheme.

The Australian Taxation Incentives for the Arts Scheme was long overdue when it was first introduced in 1978. Now, after nearly five years, it can be said that it is operating very well. It is another measure which tends towards the protection, conservation and presentation of the Australian national heritage and in particular of movables which constitute a significant part of that heritage.

These matters, including the control of illicit trafficking in cultural property, the specific protection of movables which constitute significant elements of the national heritage, and the promotion and support of public collections, are important, and warrant due consideration by individuals and Governments in Australia. A great deal of valuable work has been done by the Australian Heritage Commission and its kindred organisations (including non-Government organisations) throughout the country, but there is a need for a concerted national effort in regard to movables, which have always presented special difficulties in law and practice. (*Summer 1983*)

Andrew Lenehan — Sydney cabinetmaker

Kevin Fahy

The study and appreciation of early Australian furniture is still in its infancy. Furniture, unlike our architectural heritage, is easily portable which explains why so much has survived. More often than not divorced from its original setting, it has at least continued to serve its original function. Scattered to the four corners of the Australian continent and beyond because of the movements, depredations and additions by descendants of its original owners to their household effects, it still remains largely unrecognized for its real qualities.

In recent years its significance has been often conditioned by a buoyant nationalistic pride that takes delight in the hand-me-downs of various illustrious historic characters with little regard to its virtues other than to its most useful purpose.

Few attempts have been made to interpret what locally made furniture can tell us about the levels of colonial craftsmanship it can reflect or about the social and domestic background of the period in which it was made. Furniture is a social document in its own right that can tell us much about our past that is not always contained in conventional written documentation or printed record.

The contributions of sundry government officials, explorers, merchants and bushrangers have been more than well recorded in our historical chronicles. The story of any number of our colonial mechanics has, as yet, been hardly touched upon. But many of these men have left us tangible evidence of their work which when identified can give a new dimension to aspects of the lives and times of our forebears.

This is the story of one such man. Whether his life and work is more significant than dozens of others throughout Australia is less important than the realization that the fabric of our history is made up of more than those names usually encountered in history text books.

Andrew Lenehan (circa 1815–1886) was born in Sligo, Ireland. While the information shown on his death certificate suggests he arrived in Australia about 1832, a list of the free unassisted passengers who arrived at Sydney in August 1835 on the *Jane Goudie* includes one Andrew Lenaghan (sic).

Several billheads of Andrew Lenehan preserved in the Mitchell Library note that his business of cabinetmaker, upholsterer and undertaker was established in 1835. In May 1841 the *Australian* announced that Lenehan had acquired the cabinetmaking business of James Templeton in Castlereagh Street. Templeton was also listed among the passengers on the *Jane Goudie* that arrived at Sydney in 1835.

Although Lenehan was in business from 1841 he does not appear in Sydney directories until 1843 where he is described as a cabinetmaker at Pitt Street north and Castlereagh Street. In an 1844/5 directory his premises were listed at 271 Castlereagh Street. Joseph Fowles' *Sydney in 1848* contains an advertisement for his upholstery establishment at 287 Castlereagh Street. By 1855 he had moved to 289 Elizabeth Street and 66 Castlereagh Street. Two years later he was at 60 Castlereagh Street, more than likely a misprint, and his private residence was at *Ranelagh*, Darling Point. Between 1858 and 1866 he is listed at 179 Castlereagh Street although a billhead dated 1860 gives his address at 287 Castlereagh Street. In 1867 he appears at 181 Castlereagh Street and in the following year at the same address now described as Lenehan & Co. in partnership with J. G. Raphael and Edward Flood. An article by 'Old Chum' in the *Truth* 12 December 1915 suggests that this was an unfortunate arrangement and Lenehan vacated the premises and went to 143 Pitt Street where he is found in directories between 1869 and 1871. By 1873 he last appears at Bridge Street and within a few years he had retired to his residence on the Lane Cove River at Hunters Hill.

The varying addresses noted above are extremely useful in establishing a date for those examples of his furniture that have been found with his trade label. Apart from pieces carrying his paper trade label others are known with simply an impressed mark A. LENEHAN.

Lenehan's billhead, decorated with the Royal Coat of Arms indicating vice-regal patronage, describes him as a 'Designer and Manufacturer of Superior Furniture' which he provided for some of the finest houses in Sydney. An interesting commentary on Australian furniture is found in the *Sydney Herald* during 1846 —

Some three years ago, it was no uncommon thing to have ship after ship arriving in this colony laden with furniture, and the factories of America drew from New South Wales no inconsiderable sum for their veneered chests of drawers and painted chairs. It is pleasing to know that an example has been set in high quarters which will have the effect of discouraging the purchase of foreign or even of British furniture for our own woods and our own mechanics can now supply and manufacture articles of beauty superior even to those brought to us across the waters. The large numbers who have through the hospitality of Sir George and Lady Gipps been enabled to visit the beautiful mansion which has been erected for the dwelling of Her Majesty's representative in this colony, will have the opportunity of judging of the capabilities of the colony in this particular, and as there are many who have not eschewed the old habit of preferring imports to productions, some pains have been taken to give them a nominal description of the furniture which the Governor of the colony has thought that the colony itself could supply him; and which now sets out the festive halls, as well as the domestic apartments of the vice-regal residence. The furniture which has been added to what was in the old house, especially the massive and carved work, which has been acknowledged of considerable beauty was furnished by Mr. Lenehan of Castlereagh Street who has besides furnished some splendid articles to officials and other gentlemen in the colony.

Apart from furniture produced locally in his workshop Lenehan is known to have retailed furniture from abroad. In 1857 he advertised 'a most extensive stock of superior FURNITURE in all its variety of the most recherché and fashionable style of English and continental as well as his own manufacture'. As early as 1845 he was using not only local timbers but importing satinwood and rosewood from England for 'customers and other gentlemen who may require elegantly furnished Drawing Rooms ... he is now in the position to manufacture any article in the above-named fashionable wood, at moderate prices'.

In 1851 he was commissioned by Governor FitzRoy to make boxes from native timbers to contain specimens of gold from Ophir and the Turon which were to be presented to Queen Victoria. In a letter to Earl Grey the Governor described Lenehan as 'an ingenious and reputable cabinetmaker ... a highly respectable man, a loyal subject and member of the Corporation of Sydney'.

At the exhibition held by the Paris Exhibition Commissioners in Sydney during 1854, a chess table made of cypress pine, tulip wood and Huon pine manufactured by Andrew Lenehan was exhibited by Sir Alfred Step-

Top Billhead of Andrew Lenehan. (Courtesy of Mitchell Library)

Middle Trade label of Andrew Lenehan, c. 1845.

Bottom A red cedar breakfront bookcase with a Lenehan trade label, c. 1845, h. 252.5 cm. (Photo: A. Simpson)

Right A Brazilian rosewood occasional table with a Lenehan trade label, 1850, h. 73 cm. (Photo: A. Simpson)

hen. A later exhibition held in Sydney in 1861 by the International Exhibition Commissioners featured a cabinet made of colonial woods which was exhibited by Lenehan and obviously from his workshop.

His work continued to attract public attention and feature in the columns of the colonial press. An article on Colonial Workmanship in the *Sydney Morning Herald* during 1857 lamenting on the lack of opportunity for colonial youth to acquire skills in the higher branches of manufacturing industry noted — 'We have recently inspected a magnificent dining room suite manufactured by Mr. A. Lenehan of Castlereagh Street for His Excellency Sir William Denison, which would do honour to any cabinet factory in the world and which will provide a fitting ornament for the vice-regal dwelling'. Included is a lengthy and enthusiastic description of the suite which is still extant at Government House, Sydney, minus the mirrored backboard to the sideboard which has been dangerously relegated to the perils of Government Stores for safekeeping. Hopefully the missing member will one day be re-instated in what is a unique assembly of locally made furniture in its original, if altered, setting. The Dining Room of Government House, Sydney with its original furniture is a little known but major 19th century Australian interior.

It was certainly not Andrew Lenehan's most rewarding commission. The Votes and Proceedings of the Legislative Assembly of New South Wales in 1862 include his petition arising from a dispute over payment for this and other work he executed at Government House. The proceedings of a Select Committee to investigate his claim at least provided some belated redress and for the reader today, a wealth of information concerning the furniture and furnishing trade in Sydney about the middle of the 19th century.

Costly litigation arising from this matter could have hardly given Lenehan much comfort in his business enterprise. While local government patronage during this period was probably denied him in 1862 he is recorded as having provided furniture for Government House, Brisbane, which was first occupied in that year by Sir George Bowen who had been appointed the first governor to the newly separated colony of Queensland in 1859.

By 1875 Lenehan had retired from active business. He died at the home of his eldest son at St. Leonards predeceased by his wife and survived by three sons and two daughters. His obituary in the *Express* 4 March 1886 recorded 'Although of late years the deceased has lived a quiet and unobtrusive life, taking little interest in public affairs, yet at one time he could boast of having been foremost in almost every movement which had for its object the advancement of social and religious order. The late Mr. Lenehan was one of the first City Councillors (as far back as 1843) and for a long time took an active part in the deliberations of that august body. He could also lay claim to being one of the first of the Lay Fellows at St. John's College, an institution which he endowed to the extent of £500 in his bright and palmier days.'

Today's active concern for our architectural heritage pays scant attention to the work of cabinetmakers and other craftsmen whose labours provided many of these buildings with their original furniture and furnishings. As David Dolan wrote in an introduction to *Sydney's Colonial Craftsmen* for an exhibition held at Elizabeth Bay House by the Australiana Society in 1982 — 'The early crafts of Australia, including the crafts of Sydney, were not just a simplified provincial version of European or English or even London crafts. They were produced in an isolated society in transition, uncertain of its identity, and drawing simultaneously upon both urban and country traditions from the other side of the world, with adaptations to local materials and conditions, creating a body of work which was quite heterogeneous, and often hardly coherent, but, in that, reflecting the society in which it developed.' (*Summer 1982*)

Left A red cedar expanding dumb waiter with a Lenehan trade label, c. 1845, h. 125 cm expanded. (Photo: A. Simpson)

South West Tasmania

Josephine Flood

Dr. Josephine Flood is an archaeologist who is at present a senior conservation officer at the Australian Heritage Commission. She has undertaken extensive fieldwork on Aboriginal sites in Tasmania, N.S.W., Victoria, the A.C.T. and Queensland.

Two hundred million years ago Australia was joined to Antarctica and South America in the single great southern continent of Gondwanaland. This continent was forested with ancient tree ferns (Dicksonia antarctica) and huge trees such as the southern beech or myrtle, *Nothofagus cunninghamii*, and the Huon pine, *Dacrydium franklinii*. When Gondwanaland split up some fifty million years ago, the Australian continent gradually drifted northwards, acting as a Noah's Ark for marsupials and bearing remnants of this southern cool temperate rainforest. Now the Huon pine only survives in south-west Tasmania, where it finds the wet conditions and absence of fire it needs. Its major habitat is the Franklin–Gordon river system. There Huon pines can be found that are 2,000 years old, the oldest living trees in Australia.

The primeval rainforest of south-west Tasmania is the last great temperate wilderness in Australia, and one of the last surviving in the world. It is an outstanding example of a substantially unmodified temperate area which is of sufficient size and integrity to survive as wilderness and to permit the experience of true solitude.

There are now three national parks in south-west Tasmania: the Cradle Mountain–Lake St. Clair National Park, the Franklin Lower Gordon Wild Rivers National Park, and the Southwest National Park. These three areas total approximately 770,000 hectares in extent. They were grouped together as the Western Tasmania Wilderness National Parks and nominated to UNESCO for the World Heritage List in 1981 by the joint decision of the Tasmanian and Commonwealth Governments. The claim that south-west Tasmania is of 'outstanding universal value' was based on both the natural and cultural heritage of the region but particularly on its natural environmental values.

A region of impressive natural beauty, the South West contains unique or rare geological and biological features of immense scientific importance. Complete ecosystems remain intact, representative of pristine temperate wilderness. The flora comprises approximately 165 plant species endemic to Tasmania, of which 29 are known only in this region. The fauna includes the rare Broad-toothed Rat, the endangered White-footed Dunnart, and the Orange-bellied Parrot. The latter breeds only in south-west Tasmania, and is one of the rarest parrots in the world, with a known population of only about 200 birds.

The geological importance of the region lies in the evidence it has preserved of a major stage of the earth's evolutionary history. It contains the most glaciated area in Australia. Glacial ice has contributed to spectacular landforms in the Cradle Mountain, Lake St. Clair and Frenchman's Cap areas, carving out cirques and hundreds of glacial lakes. Wild rivers emerging from the mountains have cut across high ranges to create some awesome gorges, particularly on the Franklin and Gordon Rivers. The lower reaches of these rivers pass through limestone and dolomite outcrops giving rise to some impressive karst scenery, such as natural arches, caves and sinkholes. Karst is a scarce resource in Australia, so these features have a national importance. An even rarer feature of the

South West is the Darwin impact crater in the upper Andrew River valley, south-east of Mt. Darwin. The crater would originally have been at least 200 metres deep and 1,000 metres in diameter. Around the crater is a strewnfield of Darwin 'glass' — impact-generated glass, rich in silica and generally contorted in shape. The Darwin crater has some unique geological features which give it an international importance, and is one of the only 19 fully-established such impact craters in the world.

The high aesthetic value of the South West relates particularly to its diverse, wild, distinctive and spectacular landscape. Rugged mountain chains containing fine examples of glacial landforms alternate with alpine herbfields and broad buttongrass plains. White-water rivers roar through deep-cut gorges to the coast, which is characterized by a precipitous shoreline, long white beaches, drowned river valleys and rocky off-shore islands.

Those who put forward south-west Tasmania for the World Heritage List did so primarily on the grounds of its natural environment value, but recent archaeological discoveries have shown that its cultural importance is just as great.

In 1974 Tasmanian speleologists began to explore for caves in the limestone bluffs of the lower Gordon and Franklin Rivers. By 1979 they had discovered and recorded over a hundred caves. The possibility of traces of prehistoric human occupation existing in some of these caves was pointed out in 1977 in the South West Tasmanian Resources Survey (occasional paper) but no archaeological survey was included in the three quarters of a million dollars' worth of environmental assessment work leading to the massive *Report on the Gordon River Power Development Stage Two* (1979).

Thus the potential cultural heritage of the region was neglected, until archaeologists took the initiative and mounted two small archaeological expeditions early in 1981. The surveys were carried out in small aluminium punts and rubber dinghies, which had to be carried round a series of rapids and waterfalls. Movement on land through the dense rainforest was equally difficult, over huge old tree trunks rotten with age, through a tangle of roots and ferns and the notorious horizontal scrub. And everywhere water constantly dripped from the leafy canopy onto moss and peat on the forest floor. At times it took an hour's hard struggle to traverse as little as a hundred metres.

It is to the great credit of the archaeologists, led by Rhys Jones of the Australian National University and Don Ranson of the Tasmanian National Parks and Wildlife Service, that any human occupation sites at all were discovered in these conditions. But on the afternoon of Sunday, 11th January, 1981, Tasmania's South West gave up its first stone age secret. On the high bank of the Denison River 300 metres from its junction with the Gordon, a giant *Nothofagus* beech tree had fallen. As it was wrenched from the ground, its roots had lain bare a path of clay, revealing stone tools lying both on and in the clay deposit. Twelve tools were found, including a quartzite hammerstone. One was a quartz pebble core, with flakes chipped from it scattered around. The flakes, which would have been used as knives and scrapers, could be replaced exactly on the core, and were still as sharp as a surgeon's scalpel.

This open campsite was originally thought to be very old, but a radiocarbon date on charcoal associated with the tools gave the age of the site as approximately 300 years old. This means that even in recent times Tasmanian Aborigines at least visited this region of rainforest. The site is only 8 kilometres downstream from the famous 'First Splits' of the Gordon River. This stupendous gorge was the first seen by white men in 1928, when the party of Sticht, Harrison and Abel reached it. In the *Mercury* of 12 April 1928 the Splits are described as 'The show place of Tasmania' and it is predicted that 'they will be visited by thousands of people annually'. Little would their first white discoverers have imagined that Aborigines had penetrated this region long before.

The Denison River open campsite was the first Aboriginal site to be found in the wild rivers region, and shows that even in the last few hundred years the South West was not un-visited by Aborigines as had been thought by early ethnographers like George Augustus Robinson.

The finding of these stone tools in dense rainforest was a million to one chance, but it was followed only a month later by the discovery of rich Aboriginal cultural remains in a cave in the same region. The massive cave was originally found in 1977 by Tasmanian University geomorphology student, Kevin Kiernan, and named Fraser Cave after the Prime Minister. Kiernan revisited it in February 1981 and realised its archaeological potential. Accordingly he returned in March on a small archaeological expedition with Rhys Jones, Don Ranson, Greg Middleton and other officers of the National Parks and Wildlife Service.

Fraser Cave is located 40 metres from the east bank of the Franklin River, some 10 kilometres from its confluence with the Gordon. Its mouth is about 7 metres wide and opens directly into a large entrance chamber with a floor area of about 100 square metres. This well-lit, sheltered chamber contains most of the archaeological remains, although the cave passages extend a further 150 metres into the limestone rock.

A small pilot excavation into a bank of occupation deposit on the side of the entrance chamber revealed that this is one of the richest archaeological sites ever found in Australia. The one cubic metre of deposit excavated contained more than 80,000 stone flakes and tools and a similar quantity of animal bones. The chronology of occupation has been reconstructed from a series of radiocarbon dates obtained from layers of charcoal from ancient campfires.

The first human occupation of the cave occurred about 21,000 years ago, just before the peak of the last glacial period. This intensely cold phase 18,000 years ago caused glaciers to flow down the high mountain valleys of Tasmania, including the upper Franklin valley. Rainforest then seems to have been confined to river-edge refuges above which were open plains and alpine grasslands, similar to the arctic tundra of Alaska or Siberia. There the ice age occupants of Fraser Cave hunted wallabies and the occasional wombat and echidna. They were, it seems, highly specialized hunters. It is probable that a band of twenty to thirty people used the cave as a base camp, bringing their game back to be butchered, cooked and eaten. Many of the wallaby bones which carpet the cave floor are charred from cooking, only certain body parts are present, and long bones have been smashed open to extract the marrow.

Some knives and scrapers to fashion and sharpen spears were made from fine-grained rocks found in the glacial melt-water gravels at the river's edge. Chert and chalcedony of gemstone quality were utilized, together with an even more exotic raw material: natural glass from the Mt. Darwin impact crater some 40 kilometres to the north-west.

At the height of the last glaciation small, angular blocks of frost-shattered limestone fell from the roof to form a rubble on the floor. Nevertheless, occupation continued. Numerous interleaving hearths in the upper layer of the deposit show that the cave was used regularly for camping until about 14,500 years ago, when it was abandoned. At that time the climate was warming up, glaciers receding, and rainforest spreading upwards from its ice age riverside refuge. Whether or not this was the cause, Fraser Cave was vacated and never re-occupied. The charcoal, ash and burnt earth of old fireplaces, the charred bones and stone tools were covered with a thin layer of sterile clay and then gradually sealed in under a layer of white flowstone. Probably, for the next 14,000 years no human being entered the cave.

Fraser Cave is not unique. On the same 1981 expedition only about ten

South West Tasmania

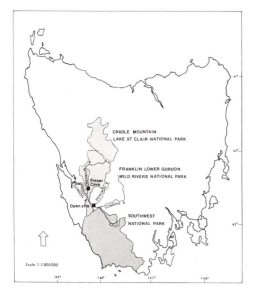

Above Lower Gordon River. (Photo: R. Turvey)

Right The Gordon River Splits. (Photo: H. Gee)

hectares of rainforest were searched, but these yielded another three definite and three potential Aboriginal sites, including another occupied cave. Then in February 1982 an expedition financed under a National Estate grant carried out a systematic search of the limestone bluffs of the lower Franklin and discovered another seven cave sites with stone tools *in situ* in deposits. One of these, with passages in excess of 300 metres, is even larger than Fraser Cave. A natural erosion gully in its floor has uncovered a rubble deposit with thin layers of old fireplaces, stone tools and burnt animal bones. Preliminary radiocarbon dates show that, like Fraser Cave, it was occupied at the height of the last glaciation. This cave (which has been given the unwieldy name of *Anadopetalum biglandulosum* after the local horizontal scrub) promises to be just as important as Fraser Cave.

Clearly the cultural significance of the South West is only just being glimpsed. Extensive karst formations containing hundreds of caves and extending over more than 100 square kilometres may well make this wild rivers region one of the richest sources of evidence concerning the late ice age anywhere in the world. Moreover, the excellent preservation conditions for bone and other organic material provided by limestone caves mean that their evidence can be used by geomorphologists, palynologists and other scientists to understand changes and processes in past and climate, flora and fauna.

To appreciate the significance of these ice age sites in south-western Tasmania, we need to see them in the context of world prehistory. The wallaby hunters living in Fraser Cave 21,000 years ago were the most southerly human beings on earth. Not for another 10,000 years would people reach equivalent latitudes in southern Patagonia and Tierra del Fuego.

In terms of their richness, age and invaluable evidence about human life at the height of the last ice age, the limestone caves of the South West can be compared with some of the classic Palaeolithic cave sites of Europe. The remains in the Franklin caves are extremely similar to those from the Dordogne in southern France. The stone tools are similar, the cooking methods are similar, even the specialized hunting strategies are similar, although northern hunters concentrated on deer and the Tasmanians on wallabies. The ice age cave paintings of sites such as Lascaux and Altamira were not done until after Fraser Cave was abandoned, but Tasmanian caves may well produce evidence of early Aboriginal art, similar to the tentative wall markings found in Koonalda Cave deep beneath the Nullarbor Plain.

The rich archaeological sites of south-west Tasmania offer significant testimony to human initiative and adaptability. It seems that almost as soon as the land link to Tasmania opened, people moved south into the new land. Before about 24,000 years ago Tasmania was an island, with the sea not far below its present level. But the world was getting colder, more and more water was being locked up in the expanding ice sheets, and sea level was dropping. By 23,000 years ago the sea had dropped over 65 metres below its modern level and the land bridge between Wilson's Promontory in Victoria and the north-east tip of Tasmania was open. Later the sea dropped still further, exposing the whole floor of what is now Bass Strait.

As soon as the gateway to Tasmania was open, people headed southwards. Why? Tasmania is further south than any other place in the southern hemisphere inhabited during the ice age. Not only were there glaciers on her icy mountains, but icebergs would have come floating past the coast from the great Antarctic ice sheet only a thousand kilometres to the south.

Into this freezing toe on the foot of the world moved the Aborigines, impelled into empty space perhaps by that same urge to explore that later drove Scott and Amundsen to the icy south. But unlike these later explorers, the Aboriginal hunters stayed and weathered out the glacial cold in caves. Cave Bay Cave on what is now Hunter Island was inhabited 23,000 years ago and a cave in the Florentine River Valley in the south central highlands was in use by 20,000 years ago. This small cave (called Beginners' Luck by its discoverers, Peter Murray from the Hobart Museum and Alfred Goede from the University of Tasmania) may well have been only used as a bivouac camp by small bands en route from the Derwent Valley to the much larger, more hospitable caves of the South West.

Both Fraser Cave and the even larger, newly discovered *Biglandulosum* Cave are incredibly rich sites. It has been estimated that there are ten million artefacts in the earth floor of Fraser Cave alone. Only one cubic metre has yet been excavated and even this has yielded scientific evidence of world significance concerning past climate, environment and human adaptation. The scientific and cultural treasures which remain to be discovered in this and other limestone caves in south-west Tasmania are incalculable.

As the furthest south ice age sites in the world, the Tasmanian caves are of inestimable importance in tracing the history of mankind. The courage and skill of these Tasmanian Aborigines, braving the ice, snow and freezing cold to hunt wallabies within sight of glaciers, is eloquent testimony to the indomitable spirit of early humans. And to the several thousand people of Aboriginal descent in Tasmania, this site is of tremendous significance as part of their heritage. (*Summer 1982*)

Australian themes in stained glass

By Beverley Sherry
with Photographs by Douglass Baglin

Formerly a Senior Lecturer in English at the University of Queensland, Beverley Sherry is now a member of the English Studies Research Centre, University of Sydney. Her interests span the visual arts and literature and she has published comparative studies of art and literature.

Australian themes in stained glass

In the history of stained glass in Australia, the years 1870 to 1914 were the period of greatest activity, a time during which stained glass became part of the general fabric of architecture, both ecclesiastic and secular.[1] Three dates of national significance fall in this period: 1870, the centenary of Captain Cook's arrival; 1888, the centenary of the first settlement; and 1901, Federation. One might expect, then, some Australian themes.

In its beginnings, stained glass in Australia was a far-flung outcrop, a colonial manifestation of the nineteenth-century revival of stained glass in Europe, a movement in which England led the way.[2] The influence of the 'mother country' was fundamental and artists working in Australia regarded themselves as participating in the general nineteenth-century rediscovery of stained glass. The most influential of the early stained-glass artists and firms in this country, such as Lyon & Cottier in Sydney and William Montgomery in Melbourne, were European-trained. Montgomery (1850–1927) is typical: having worked for the English firm of Clayton & Bell, he gained seven years further experience in Munich, then

Left Australia window, Sydney Town Hall, designed by Lucien Henry and manufactured by Goodlet & Smith, 1889.

Above Rose window, St Bede's Church of England, Drummoyne, Sydney, 1931. A new treatment of an ancient form, with Australian wildflowers forming the rose.

Below Kookaburra on the fanlight of the front door in a house in Hunter's Hill, Sydney, 1895.

Top right Captain Cook window, Sydney Town Hall, designed by Lucien Henry and manufactured by Goodlet & Smith in 1889.

Far right West window, St Augustine's Church of England, Hamilton, Brisbane, designed by William Bustard and manufactured by R. S. Exton & Co. in 1948.

Right Waratahs in a staircase window, St Cloud, Burwood, Sydney, 1893.

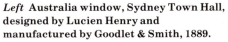

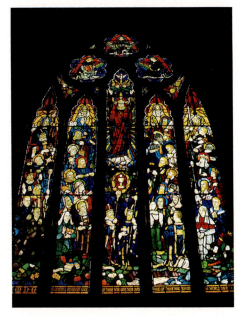

came to Australia in 1887; he opened a studio in Flinders Street, Melbourne, soon had samples of his work throughout the country, and in his lectures and publications taught the revived principles of stained glass, just as Charles Winston did in England.[3]

With ecclesiastical glass, it was natural that artists in Australia should follow the old world, since the ecclesiastic was a venerable tradition with set biblical themes and iconography. In secular stained glass too the English influence was strong. Pre-Raphaelitism is evident in the stained glass of late nineteenth-century and early twentieth-century houses and public buildings in Australia, specifically in themes from medieval romance, British history, allegories of the arts, characters from Shakespeare, and the ubiquitous four Seasons. Admittedly there was a national pride in the achievement of local artists, an admiration for the technical manipulation of glass 'to suit the high lights of our colonies,' and a confident attitude that the young country could produce glass equal to any from Europe.[4] But did the early artists attempt to adapt their art imaginatively to the new environment and find new themes for the new land?

In the churches, there were few Australian themes in the period 1870–1914. Understandably, it was the two World Wars that inspired such themes, in memorial windows. The Victorian habit of idealising the dead by representing them as knights in armour persisted well into the twentieth century, but there appeared also, in some windows, figures in the plain uniforms of the various Australian forces. A First World War memorial in The King's School Chapel, North Parramatta, N.S.W., depicts three soldiers in the khaki uniform and slouched hat of the A.I.F. and, in the background, a building representing The King's School; the memorial is to Stewart Milson, an old boy killed at Gallipoli in 1915, and the stained glass is the work of Sydney artist, Norman Carter (1875–1963). The Queensland artist, William Bustard (1894–1973), who migrated to Australia in 1921, designed several such windows for the Second World War, the most ambitious being the *Te Deum* of St. Augustine's, Hamilton, Brisbane, in which a soldier, sailor, and airman, as well as three Australians regarded as New Guinea martyrs, are placed in a company of saints and martyrs. With the increasing secularization of the churches in the twentieth century, other Australian themes have appeared: the work of the Australian Inland Mission is represented in stained glass in the Uniting Church, Toowong, Brisbane; the early settlement of Gippsland is recorded in a window of the Cathedral Church of St. Paul, Sale, Victoria; and sometimes Australian floral emblems appear, as in the unusual rose window of St. Bede's, Drummoyne, Sydney.

Australian themes inevitably found fuller and freer expression, though, in secular stained glass — in both houses and public buildings.

From the time of the centenary of 1888 up to Federation in 1901, stained glass was a regular feature of domestic architecture. Familiar motifs in the fanlights of front doors were Australian flowers or birds. Sometimes such themes were treated more grandly in staircase windows. When W. C. Payne built Studley Park at Narellan, N.S.W., in 1888, he had a tower window designed to order in which his family coat of arms was combined with waratahs;[5] in 1893, George Hoskins, father of the Australian steel industry, made the focal point of the entrance hall of his home a staircase window depicting a vase of waratahs. In the year of the 1888 centenary, William Montgomery designed an exemplary set of windows with Australian fauna and flora for Sir William Clarke's residence, Cliveden, in Melbourne,[6] but they disappeared when Cliveden was demolished; ironically, one of Montgomery's most English windows, the medieval hunting scene in Tay Creggan, Hawthorn, Melbourne, has survived unscathed.

Two houses of special interest for their Australian themes are Yanco (now Yanco Agricultural High School), Yanco, N.S.W., and Cranbrook (now Cranbrook School), Bellevue Hill, Sydney.

The glass at Yanco is signed and dated and was made by the Sydney firm of F. Ashwin & Co. in 1902. Frederick Ashwin (c.1835–1909) came to Australia from Birmingham in the early 1870s and went into partnership with Sydney's first stained-glass artist, John Falconer (c.1838–1891); some of their earliest work (1877) is the stained-glass dome in the vestibule, the first part, of the Sydney Town Hall.[7] After Falconer's death, Ashwin continued independently as F. Ashwin & Co. The stained glass at Yanco shows a transition from old themes to new: an Arcadian shepherdess is attended by a Cupid playing a pipe, but her flock is clearly identifiable as merinos. Yanco was completed in 1902 for one of Australia's leading pastoralists, Samuel McCaughey; then sixty-seven years old, McCaughey had spent much of his life trying to improve the quality of Australian wool and was knighted for his efforts in 1905. The stained glass is a tribute to him in art and depicts what seem to be prize specimens of merinos. The artist has adapted traditional themes — the shepherdess, the Cupid — to the context of a particular house and its owner and to the themes of the new land.

The stained glass at Cranbrook, a set of 'Captain Cook' panels, is more ambitious. Made by the Sydney firm of Lyon & Cottier,[8] the window was installed after the Hon. James White bought Cranbrook and had it remodelled in 1874. It is no doubt connected with centenary celebrations for Cook's voyages of the 1770s; another 'Captain Cook' window, with a border of Australian fauna and flora, was exhibited by Lyon & Cottier at the Melbourne Exhibition of 1875 and later installed in the staircase of Lyon's home.[9] John Lamb Lyon (1835–1916) and Daniel Cottier (1838–1891) had trained together in Glasgow, an important centre for stained glass, and their Sydney firm opened in 1873; since Cottier did not live permanently in Australia, the firm was run principally by Lyon. While much of their work is Pre-Raphaelite in character, the Cranbrook window represents a strong departure — the writer's experience of discovering it the same day as seeing Lyon & Cottier's very English 'St. George' window at the Waverley War Memorial Hospital in Sydney (the Vickery home of the 1880s) was refreshing to say the least.

Lyon was a recognised portrait painter as well as a stained-glass artist and his dual abilities are manifest at Cranbrook. Captain Cook and members of his expedition, including the naturalists, Banks and Solander, are portrayed. Cook's journal makes repeated reference to the kangaroo, the animal 'which had been so much the subject of our speculation,'[10] and in the centre panel Lyon depicts the first sight of a kangaroo, with two men looking mystified and startled. History is recorded in these panels dynamically and with realism; in every scene, Lyon has included authentic details of setting and has captured physical movement and human reactions. Purists may condemn the window for trying to be what it is not, a painting, but it may be regarded also as an interesting sample of the hybrid work of an artist who has dual talents, those of portrait painter and glazier.

Houses speak for individuals and families, public buildings and institutions for larger groups, thus offering further scope for Australian themes. In public buildings, Australian flora and fauna appear: the waratahs in Sydney's Central Railway Station (1906) are signed as the work of the Sydney firm of Goodlet & Smith, as are the waratahs and flannel flowers in Sydney Hospital (1893); also signed is the glass in Parliament House, Adelaide, made in 1888 by Smyrk & Rogers of Melbourne and incorporating Australian flowers. The Australian coat of arms was also commonly used, as in the domes of the Supreme Court Library in Melbourne (1884) and the Westpac Bank, 228 Pitt Street, Sydney (1912). More expressive, though, are windows which record history or depict industries. The Commonwealth Savings Bank in Martin Place, Sydney (1928), has a Grand

Top Sower and Reaper, Land Administration Building, Brisbane, by George Gough, 1905.
Above Details from Captain Cook window, Cranbrook, Bellevue Hill, Sydney, by Lyon & Cottier, c 1874. Nine panels make up the window. The centre panel shows Mr Banks lecturing Dr Solander. The kangaroo proved difficult for early artists and Lyon's attempt (*right*) is not very successful.

Right Pastoral scene, Yanco, N.S.W., by F. Ashwin & Co., 1902.

Australian themes in stained glass

Hall surmounted by a vaulted ceiling with stained-glass panels showing scenes representing the sources of the nation's wealth — agriculture, grazing, shipping, building, mining; the panels were manufactured by John Ashwin & Co.[11] Similar in concept is the stained glass in the Land Administration Building in Brisbane (1905), in which figures from rural life, in keeping with the function of the building, are represented — Sower, Reaper, Squatter, Pioneer, Tiller, Herdsman; this is the work of one of Queensland's first stained-glass artists, George Gough.[12]

In 1901, Federation inspired one of the most important examples of stained glass in public buildings, the three-light window of the Adelaide Stock Exchange. Britannia is portrayed in the centre light, looking like the Roman goddess Minerva and attended by figures representing Australia (holding a jewel box), India, Africa, and Canada; three smaller lights above represent Morning, the Sun, and Evening ('the sun never sets on the British Empire'). Britannia holds a gold nugget set in a laurel wreath and inscribed, 'Federation'; presumably the gold is a gift from Australia and not inappropriate to a Stock Exchange and to the donor of the window, the Hon. George Brookman, who made a fortune from the

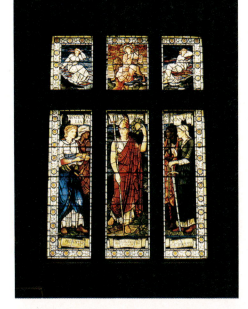

Above centre Federation window, Adelaide Stock Exchange, designed by John Henry Dearle and manufactured by Morris & Co. in 1902. (Photo: K. L. Goodwin)

Above right Australia window, Melbourne Town Hall, 1888.

Right Pioneer, Land Administration Building, Brisbane, by George Gough, 1905. Gough's original panel of the Squatter was lost but replaced in 1979 to a design by Oliver Cowley and manufactured by Brisbane Leadlight Service.

73

Kalgoorlie gold fields.[13] The window was designed by the English artist, John Henry Dearle, manufactured by the English firm of Morris & Co., and is an attempt to adapt the Pre-Raphaelite allegorical manner, characteristic of Morris & Co., to an Australian theme.[14]

It was the 1888 centenary, however, that inspired what is probably the most outstanding example of Australian themes, the stained glass in the Sydney Town Hall. The main hall itself was built for the centenary and an important part of its design was the stained glass installed in the clerestory of the auditorium and in the staircases leading to the galleries. The clerestory glass, made by Goodlet & Smith,[15] depicts Australian fruits, crops, and flowers. More powerful expressions of nationalism, though, are the staircase windows, manufactured by Goodlet & Smith and designed by Lucien Henry (1850–1896).[16] Henry had studied under Viollet-le-Duc, leader of the stained-glass revival in France, and he lived in Australia from 1879 to 1891 and pioneered the introduction of Australian flora and fauna into art. In strong, glowing colours, he portrays Captain Cook in the northern staircase and a female figure representing Australia in, appropriately, the southern. The Australia window is replete with iconography: in the fleece and ram's horns we are to read the wool industry, the miner's lamp mineral wealth, the trident maritime power, the Union Jack our British origins, the five stars of the Southern Cross and the globe inscribed 'Oceania' our position in the world; the border of the window is filled with splendid examples of Australian wild flowers — waratah, flannel flower, and stenocarpus (fire-wheel). An equivalent window was made by an Austrian firm for the Melbourne Town Hall but it lacks the fervour, dramatic impact, and breadth of meaning of Henry's windows.

Some artists, then, brought to Australia the benefit of their European training and, while drawing strength from that training, adapted their art imaginatively to the new environment. Foremost were John Lyon with the Cranbrook windows and Lucien Henry with the Sydney Town Hall windows. All the windows with Australian themes, however, express social and national awareness; in different ways and with varying artistic success, they express the feelings of individuals, families, institutions or groups — all the windows are saying something. There is at present a revival of interest in stained glass in Australia and some of our artists are reaching out to communicate with the world of stained glass in Europe and the United States; such communication was a strength of the early artists. Considering this, may we hope that architecture will offer the opportunities and the bicentenary of 1988 perhaps the inspiration for stained-glass artists to say something again. (*Summer 1983*)

War memorials in our landscape

K. S. Inglis

K. S. Inglis is Professor of History at the Research School of Social Sciences, Australian National University. He has published several books and is currently making a systematic study and analysis of war memorials in Australia.

D. H. Lawrence, exploring the alien Australian landscape in 1922, noticed in the little town of Thirroul, north of Wollongong, a memorial to the men who had gone to the Great War. In his novel *Kangaroo*, where Thirroul became 'Mullumbimby', Lawrence described the object as 'small and stiff and rather touching. The pedestal was in very nice proportion, and had at eye level white inlet slabs between little columns of grey granite, bearing the

Left C. Web Gilbert's War Memorial, Burnside, S.A. (Photo: K. S. Inglis)

Above Margaret Baskerville's bronze soldier, Maryborough, Vic. (From *Margaret Baskerville, Sculptor.* Melbourne, 1929)

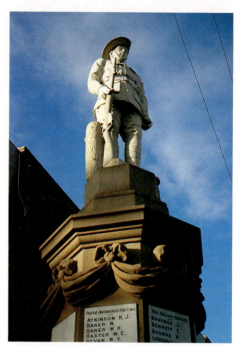

Right War Memorial, Ulverstone, Tas. (Photo: M. Bourke)

Far right World War I Memorial, Thirroul, N.S.W. (Photo: K. S. Inglis)

names of the fallen on one slab, in small block letters, and on the other slabs the names of all the men who served: "God Bless Them". The fallen had "Lest we forget" for a motto. Carved on the bottom step it said, "Unveiled by Granny Rhys". A real township monument, bearing the names of everybody possible: the fallen, all those who donned khaki, the people who presented it, and Granny Rhys. Wonderfully in keeping with the place and its people, naive but quite attractive, with the stiff, pallid, delicate fawn-coloured soldier standing forever stiff and pathetic.'

'Standing forever' — that was prescient of Lawrence, observing the figure with slouch hat, bowed head and reversed rifle when the Digger had been on his pedestal for only two years. Six decades later he may still be seen. He is not exactly where he was in 1922, for like many of his kind he yielded to progress after becoming a hazard to motor traffic; and he may be moved again soon because the R.S.L. Club he guards has gone bankrupt and the land in which he is planted may have to be sold. But fresh paint, picking out his belt and puttees and hat, giving his face an embalmed look, shows that he is cherished. So does the history of him unearthed from the pages of the *South Coast Times* by Ted Johnson, president of the Thirroul R.S.L. sub-branch, and issued in 1977.

Women of Thirroul began to collect money for a monument while their men were still fighting and dying. Grannie Riach (Lawrence's 'Granny Rhys') and Mrs. Arnold Higgins collected money from coal miners and railway workers and their families; raffles and dances and jumble sales raised more. Nearly two hundred pounds was in the fund by the time the surviving soldiers came home. Returned men and the relatives of dead men were asked to supply the post-master with names for the slabs. An architect from Wollongong and a local builder gave their services free, and so did a plumber from Bulli, for the monument was conceived as a fountain, and it was the Soldiers' Memorial Fountain whose foundation stone Mrs. Higgins declared well and truly laid on 24 October 1919. By Anzac Day 1920 it had become the Memorial Monument and Fountain. The town turned out for a ceremony that day. 'As the members of the procession were taking their places around the enclosure,' wrote a delighted reporter, 'it was noticed that a kookaburra (the laughing jack-ass), that grand quaint bird so typical of Australia, had alighted on an electric light pole overlooking the scene as if to symbolize Nature's approval . . .' Grannie Riach unveiled the white freestone figure carved by Mr. Casagrandi, of Hurstville, and said that the memorial would 'serve to remind everyone of our brave heroes who kept Australia safe from the horrors of warfare'. 'I hope', the old lady went on, 'that the people of Thirroul will see that it is kept as a sacred memorial to our brave soldiers and sailors of Thirroul.'

Thirroul was the first town on the south coast of N.S.W. to put up such a memorial, and inspired neighbouring places to make their own. Before long Bulli-Woonona had a tall column topped by a sphere, Austinmer a pedestal topped by nothing, Wollongong a fat arch with a naked figure of Victory squatting on it (the unpaid design of Varney Parkes, son of Sir Henry). All across the continent municipal councils and R.S.L. sub-branches and ad hoc committees put up monuments to record the names of their men and to be sacred sites, places where wreaths were laid and people gathered on the proper occasions, especially on Anzac Day at dawn services and after marches later in the morning.

The war memorial was there to help people express and assuage their grief. It was there to let the civic leaders who had sent men to their death provide reassurance in symbol and word that those deaths had meaning, that men had died for country and empire and freedom and justice and humanity, sometimes for God. It was there to entreat remembrance from the living and from posterity.

Some places already had war memorials in 1914, to men who had served the British empire in South Africa between 1899 and 1902. But only 16,000 men had gone to that war, and so few died there — about 500 — that some of the monuments were to individuals. That was not feasible after 1914–18, when more than 60,000 died. About one citizen in every seven went away to the Great War — one in three or four of the males; every second man between 18 and 45. More than half of the men who returned had suffered some sort of wound. To the average community, between one in five and one in six of the men who went away on the troopships did not return. On most memorials a symbol against a man's name indicated that he had died at the war. If you count, you will discover how evenly the killing of young males was spread across the nation. None of their bodies (except one, General Bridges') was shipped home, and few Australian mourners could expect to visit the white stone cemeteries of the Imperial War Graves Commission in Europe and the Middle East. American families, by contrast, could have their dead sons buried at home if they chose; and English people could at least think seriously about visiting graves in foreign fields across the Channel. In Australia, the

Above War Memorial, Port Germein, S.A. (Photo: B. Gammage)

Left Boer War Memorial, Brisbane. (Photo: K. S. Inglis)

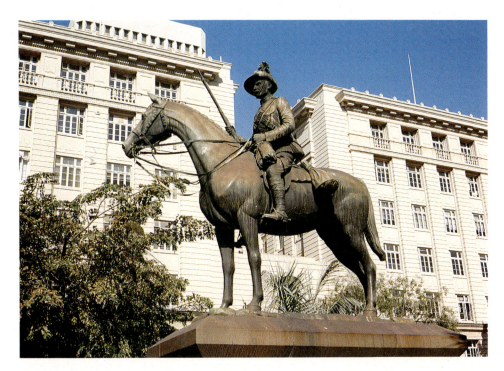

burial of so many loved ones on the other side of the world enhanced the meaning of a war memorial: looking at a name on the pedestal, placing a wreath against the base, could be a substitute for visiting a grave. The newly current word *cenotaph* was much used by Australians, even when the memorial described was not what that word means, an empty tomb. Whatever its design, an empty tomb is what many people felt a war memorial to be.

With or without advice from architects and sculptors and clergymen and other experts, communities chose what seemed to them the right design. The soldier carved in stone was probably the most popular choice. The artist Arthur Horner, depicting in *The Age* Myths and Monsters of Down Under, has among his exhibits 'the Anzac: a brave youth who fought the enemies of his master, and was turned into stone on his return'. In Cairns and in Port Germein (S.A.) he stands high on a pedestal. In Burra (S.A.) he stands even higher on a column, bayonet up, coming about as close as a stone figure can to lunging. At Murrumburrah (N.S.W.) he has the face of an angel. In a few places the soldier is cast in bronze, the work of a professional sculptor. C. Web Gilbert's in Burnside (S.A.), is perhaps the most athletic soldier-figure in Australia (and Gilbert's best work, in the estimate of another sculptor, John Dowie, who sees it as 'heroic, but with something of the stink and terror of the trenches about it'). Margaret Baskerville's bronze soldier in Maryborough (Vic.) is, I think, the only war memorial in Australia made by a woman (and was commissioned by the local Women's Patriotic League). The other Maryborough (Q.), has not only a soldier but a sailor, an airman and a nurse, each in stone on a pedestal and all surmounted by an angel with sword aloft.

But there are many other motifs, and a census may well show that soldier-figures are in a minority. At Mildura (Vic.) a female figure mourns beside a cross. At Loxton (S.A.) the soldiers are commemorated by a rotunda. Dubbo gives them a great tower. Port Pirie (S.A.) has gates at the entrance of a memorial park. Griffith (N.S.W.) has a tapering tower in Memorial Gardens even though the town did not exist in 1914–18: the people who made Griffith in the 1920s, like Australians everywhere else, needed their war memorial.

By 1930 the landscape was full of them. In the 1930s came the grand 'national' memorials in each capital city. When the Great War had to be re-named World War I, each community had to decide whether to attach new names to the old monument or to make a fresh tribute. If they did opt for a separate World War II memorial it was likely to be 'useful', having some secular as well as sacred purpose: a hospital, a hall, a swimming pool. The people of Ulverstone (Tas.) were unusual in commissioning an edifice more elaborate than the original, enclosing the Digger of the 1920s within three concrete piers supporting a four-faced clock, on top of which a pinnacle lights up at night. The artist George Sprod, a fine observer of the genre, described this one in *Quadrant*, October 1975: 'Replete with imagery the monument rises from a stone map of Tasmania, coloured blue, and a round base representing the sea. The columns are in the form of elongated books, the pages represented by glass panels enclosing the names of those who served. These books are anchored together by massive chains, the chains of comradeship.' What would D. H. Lawrence have made of it?

Today it seems unlikely that a construction such as Ulverstone's will ever inspire the sentiment evoked in Lawrence by that figure at Thirroul. But tastes change. The stone soldiers were often derided when they were new. Another war might flatten the monuments to earlier conflicts and leave no mourners. Short of that, all our war memorials could well become more precious as they age, to be cherished collectively as part of the national estate. (*Summer 1983*)

Above and right Details of War Memorial, Maryborough, Qld. (Photos: E. Waters)

Cemeteries — their value and conservation

James Semple Kerr

Dr. J. S. Kerr is a former Assistant Director of both the National Trust of Australia (N.S.W.) and of the Australian Heritage Commission. He is at present Chairman of the Historic Buildings Committee of the Trust and a member of the Technical Advisory Committee of the Commission.

During the last decade cemeteries have come under increasingly destructive pressure. It has involved alienation of land, conversion to other uses, ill-conceived 'restoration' and landscaping, short-cut maintenance and, surprisingly, a failure to plan for and control a resurgence in the construction of new monuments. Simple neglect over the last half century has had not only a slow destructive effect on physical fabrics but it has led insidiously to a municipal belief that cemeteries are untidy, worthless places ripe for more productive uses.

Almost all our present problems with cemeteries stem from a failure to understand their value to our community. It therefore seems an appropriate time to re-examine some aspects of their significance and to look briefly at the way future decisions might be approached.

Most people are already aware, through the work of historical societies and genealogists, that cemeteries are useful as a source of biographical information and that they are likely to have strong associational significance for the local inhabitants. However, other aspects contribute to their significance and may be less well understood. They relate to the capacity of the place to explain or demonstrate past practices and to its quality.

This capacity of the place to demonstrate a custom, taste or function is of particular importance as cemeteries often provide the least altered physical evidence of our changing cultural attitudes. The burial place at Kingston, Norfolk Island, despite some recent rearrangement of monuments, demonstrates clearly what an early Australian colonial cemetery in the Georgian tradition was like. Tomb-chests and a range of single-plane headstones with simple silhouettes and carving restricted to low relief were typical. Decorative motifs were taken from the contemporary vocabulary of Neo-classicism. Lettering was mostly in a handsome Roman serif face or script.

The layout was in roughly parallel rows and the orientation easterly, but not rigidly so. The early stones in such burial places occasionally reflect the seasonal variations of the rising sun — a compass not being the usual equipment of pioneer grave-diggers and masons. Some of the early country burial grounds remained unsurveyed and undedicated for many years, and the informality of the early stones is in sharp contrast with the rigid rectilinearity of the post survey layout. The earliest stones at Dubbo General and Goulburn Church of England are examples.

The English tradition of churchyard burials survived with decreasing frequency throughout the nineteenth century. Cobbity, N.S.W. is an early and intimate example and Hagley in Tasmania is a later representative with distinct landmark quality.

However, by the mid nineteenth century the typical country town

Left Fraser Mausoleum, Rookwood, N.S.W. (Photo: J. S. Kerr)

cemetery was designed or adapted to be a general cemetery. It was surveyed and had a central avenue or parallel avenues which served the separate denominational areas. It was rectangular and the graves were laid out according to an internal rectilinear grid, very much like contemporary town planning. It was usually surrounded by a perimeter plantation.

In the latter half of the century a few grand metropolitan cemeteries like Melbourne General, Rookwood and Waverley were laid out according to Picturesque and occasionally Gardenesque principles then in vogue in England and America. Between the serpentine walks the usual rectilinear grid was fitted with some difficulty. Whether churchyard, country general or metropolitan Picturesque, plantings played an important part in reinforcing the cemetery plan particularly in the second half of the century. Dark evergreens were favoured and the choice had a symbolic significance as well as landscape importance.

In addition to the demonstrative capacity of layout and planting, the memorials themselves illustrate social and religious attitudes, popular imagery and even the architectural tastes of their day. In iconographic terms the headstones of the nineteenth century are of extraordinary interest and, together with stylistic characteristics, evolved throughout the period.

In the oldest colony, N.S.W., the century started with the survival of a few genuine and more or less doctrinally correct Christian medieval icons. The headstone for Bridget Egan at Parramatta is the earliest known example. It is one of the standard representations of the crucifixion — Christ with his crown of thorns, his hands and sides dropping gouts of blood with the sun and St. John (the disciple whom Jesus loved) on his left and the moon and the Blessed Virgin Mary on his right. This was not a self-conscious revival, it was the end of a very long tradition. Variants

Top Sutton Forest Roman Catholic Churchyard, N.S.W.

Above Wagga Wagga lawn cemetery, N.S.W.

Right Richmond, Tasmania, view from the cemetery. (Photos: J. S. Kerr)

Above right Kingston burial ground, Norfolk Island. (Photo: E. J. Kerr)

Above Headstone of F. A. Bunbury, 1858, Botany ex Devonshire St, Sydney. (Photo: J. S. Kerr)

can be found at St. John's burial ground at Parramatta and at Berrima, Raymond Terrace and Port Macquarie, all in N.S.W.

By the 1850s I know of no such survivors. Instead by the end of the decade Victorian story-telling images were being carved in the round or in high relief. The themes of faith, hope for the resurrection and the ritual of bereavement were still there, but where the early carving was formal and symbolic with implicit meanings, it was now often mundane, explicit and easily understood. The destination of Florence Angelina Bunbury's soul in 1858, and the authority for this information, is made angelically clear on her memorial.

During the 1860s High Victorian architectural motifs, particularly Gothic, became common. John Gibsons' memorials to the Molyneaux family and to himself, both dating from 1866 and in Kelsall Green Cemetery, London, are grand examples. The Roche family memorials from 1866 to 1973 at Yass, N.S.W. exhibit similar motifs — gables, sometimes triple, supported by colonnettes with bases and capitals or by buttresses. It is of interest that the stylistic and iconographic development of monuments in the Australian colonies does not support any time lag theory. N.S.W. for example is neither less nor more advanced than a major provincial area in England.

By the 1880s advances in marketing techniques were creating a fusion of popular art and religious sentiment. This reached its apogee in the first decade of the twentieth century with the near universal popularity of touring great works of art, as well as prints, oleographs, lantern slides and postcards. Many of the images concerned faith and death and found contemporary expression in monumental masonry and not a few were related to popular songs or hymns.

One example will suffice. Readers may recall the story of the Reverend Augustus Montague Toplady sheltering from the rain under a large cleft rock in Burrington Gorge, Somerset and of his subsequent writing of the hymn *Rock of Ages*. It became immensely popular and Selbourne noted in the 9th Edition (1880s) of the *Encyclopaedia Britannica* that, 'some think it the finest in the English language'.

Probably in the early 1870s it inspired a series of paintings which became compelling and popular images, such as J. A. Oertel's version. Another was by Gustave Lenori. An advertisement in the *Town & Country Journal* of 1878 gives an idea of the market penetration of these works.

The Imperial Art Association's Grand Premium Gift, for 90 days only, to be sent to every reader of this Paper, consisting of the valuable loaded oleograph, entitled,

'ROCK OF AGES'
Size 24 × 32 inches

. . . the only exact copy of Gustave Lenori's celebrated painting, and which was sold at the enormous price of 30,000 dollars.

Between the 1890s and the 1930s several versions of the picture appeared on a whole range of postcards and chromolithographs. The image had appeared on monumental masonry by the early 1880s but it did not reach its definitive and most popular form until just after the turn of the century. It then remained a favourite for twenty-five years.

Mr. Shakespeare, a monumental mason of Wellington, had an example carved as a memorial to his thirteen year old daughter in 1912. The treatment of dress and face suggests it was a portrait of the girl, although most country masons ordered similar works from metropolitan or overseas catalogues. A marble catalogue from Vermont, U.S.A., confirms that the *Rock of Ages* design was an international phenomenon.

At the other end of the scale to these highly developed manifestations

Headstone for Kathleen Shakespeare, 1912. Wellington General. (Photo: J. S. Kerr)

of popular culture were the anonymous but extremely evocative burial places of those with neither money nor appropriate skills. The mounds of rock with their rough uninscribed headstones behind the Explorer's Tree on Pulpit Hill near Katoomba are traditionally thought to have been graves of early road gang convicts, although this provenance has not yet been fully established.

In country areas the graves of children are often set with the child's favourite toy. A silver painted horse at Kempsey, a much sucked terra cotta doll at Tenterfield, a meccano set at Yass and a diminutive and hardly used pair of copper-toed boots at Boorowa are typical.

However, the most moving and iconographically pure is perhaps an unidentified grave at Tenterfield. It has a cross as a symbol of faith and a border, both carried out in shells — a traditional material associated with burials for at least 3,000 years. At the head is a lamp to light the way according to Christ's promise: 'I am the light of the world: he that followeth me shall not walk in darkness ...'. Last there is a sturdy hardwood peg driven into the ground so that the location is marked even if surface elements are obliterated. It is a complete statement without words by someone with faith, hope and charity, but no money. Who shall decide that such memorials are sufficiently unimportant to be bulldozed to make way for a 'landscaped rest park'?

In addition to the fact that cemeteries illustrate past ideas of planning, iconography and stylistic development, they are a major repository of craft skills in stone, iron and to some extent timber. They are in effect an open air museum revealing otherwise unknown craftsmen and their work.

From such physical evidence it is possible to distinguish the work of some 350 monumental masons in N.S.W. during the nineteenth century. They are of variable quality but all are of interest. An examination of their monuments reveals the itinerant characters of many masons, their sources of inspiration, their technical abilities as well as their use of local materials and in some areas the rise and decline of regional characteristics and family dynasties.

While it should be remembered that cemeteries were designed to be receptacles for the dead or their ashes, they were also intended to provide a means by which the living — relatives, friends, descendants and visitors — could continue to honour their memory. Hence our Victorian and Edwardian forefathers have already installed most of the necessary amenities for presentation, interpretation and public usage. There are usually plans showing denominational areas, signs giving directions, necessary drainage, drives and paths for access, landscaping and plantings to create a beautiful and restful environment secluded from the world, and fencing to protect the monuments and plantings from livestock. In the larger cemeteries shelters, chapels and public conveniences were erected. In short, cemeteries are, by design and intention, suited to public use.

In most cases what is now necessary is tender care and modest conservation to repair the damage to existing facilities and features from half a century of neglect and, in metropolitan areas particularly, destruction. Some adaptation, reconstruction, restoration and new construction may be necessary, but only if the significance of the place is first understood and the new work preserves the capacity of the place to demonstrate that significance. The principles guiding such work are set out in the Australia ICOMOS Charter for the Conservation of Places of Cultural Significance, commonly known as the Burra Charter.

A second and subsidiary way in which cemeteries may be significant, is in their degree of landscape or environmental value. This category is to do with quality and excellence. At its most subjective we are simply asking ourselves, is it a pleasant place to be in and, if so, why?

I have separated this quality because cemeteries may well demonstrate some significant aspect of history and be comparatively unattractive, just as the environmental attractions of other cemeteries may be largely fortuitous. However, such cases are exceptional. Usually qualities of excellence reinforce the capacity of the place to demonstrate some aspect of the past and enhance its value.

Why then are some cemeteries pleasing? The answers are much the same for any aspect of the built environment. Firstly because their elements possess a degree of unity in their material, form and scale. Take the examples of Helmsley, Yorkshire and Sutton Forest Roman Catholic Churchyard. Clearly the consistent use of materials of similar texture and colour, the application of consonant forms, and the human scale, make a most agreeable visual setting.

All these examples exhibit some variety of form and scale within a restricted range. This helps to avoid monotony. However it must be stressed that historic significance is much more important than aesthetic improvements. Hence the 'monotony' of a war cemetery or a monastic or conventual burial plot results from a concept of the utmost historical significance — that all people are equal in the sight of God. It would be absurd to attempt to tamper with such arrangements. Places with a considerable degree of variety are given a greater degree of coherence by a common element such as the awnings and posts of a town like Chiltern in Victoria or the ornamental cast-iron railings of St. Jude's cemetery, Randwick, N.S.W.

Where all the unities of materials, form and scale are violated the results are disastrous. Australia Post's ill mannered intrusion between the red brick Gothic towers of the school and the Presbyterian Church in Howick Street, Bathurst is nicely paralleled by the recent erection of two large textured brick mausolea in the middle of one of the most harmonious small scale burial grounds across the Blue Mountains — at Hartley.

Above Roche family headstones, 1864–73, Yass, N.S.W. (Photo: J. S. Kerr)

Right Helmsley, Yorkshire.

Bottom Headstone for Bridget Egan, Parramatta.

Middle Pulpit Hill burial ground, near Katoomba.

Below Hartley cemetery, N.S.W. (Photos: J. S. Kerr)

The three unities *can* be broken, but one at a time and then only with thoughtful attention to siting and to the treatment of the two remaining unities. The result can be dramatically and symbolically effective as in the Fraser Mausoleum at Rookwood.

The second characteristic which may enhance the landscape quality of a cemetery is its relationship with its setting. At Richmond in Tasmania the Church of England cemetery has a gentle but direct visual link across the river with the church, the town and its famous sandstone bridge. The Catholic cemetery by contrast is dramatically sited past the church on a narrow ridge with an escarpment on each side and terminating in a bluff above the Coal River. The location of Waverley cemetery, overlooking the expanse of the Pacific Ocean, is even grander.

Many country cemeteries have a more subtle but equally satisfying relationship with their setting. Collits' is entirely enclosed in bushland near the foot of the Mount York escarpment, whereas Boorowa sits astride a low saddle amid lightly timbered rolling hills. At neither place has any form of landscaping been attempted and both are pleasing because of the care with which their location has been chosen and because of the absence of distracting intrusions.

The third aspect of environmental quality relates to landscaping and plantings within the cemetery itself. Most country and metropolitan cemeteries designed in the latter part of the nineteenth century were intended to have a perimeter plantation. It offered practical advantages as well as symbolic seclusion. It excluded visual distractions, acted as a windbreak in winter and gave some relief from the heat of summer.

In addition the internal plan of the cemetery was reinforced by plantings, mostly formal, of pines, cypresses, palms and, particularly in the drier areas, kurrajongs. Almost all these cemeteries are now landscapes in distress and are in urgent need of a replanting programme. Over the

Australia in Trust

last fifteen years the untutored use of non-selective herbicides as an alternative to normal maintenance has decimated cemetery plantings. The original Stuart cemetery at Alice Springs has recently lost its lovely, slow growing corkwoods and the central avenue at Dubbo now consists of stumps, gaps and oleander. Where replanting programmes have been carried out by local government, quick growing natives or showy exotics have usually been preferred to the traditional and slower maturing evergreens which our Victorian and Edwardian forefathers planted for posterity.

Cemeteries are historical documents illuminating a whole range of historical themes. They not only help explain our past but provide a sense of continuity and identity. In addition they have usually been, sometimes are, and certainly can be, places in which it is pleasant to walk abroad and recreate oneself.

Since the 1920s there has been an accelerating decline in the quality of cemetery and memorial design and maintenance which has accompanied an increasing desire to avoid involvement in the process and ritual of death. This has led with inevitable logic to proposals for the conversion or recycling of cemeteries and to the proliferation of new cemeteries in which the over-riding design criterion is minimum maintenance. Wagga Wagga lawn cemetery is a well kept example. The twenty-first century may not place it high on the scale of twentieth century achievements but it will certainly regard the type as a telling demonstration of our municipal aesthetics.

However, the encouraging sign is that we do appear to have passed the low point of community indifference. People are starting to take a greater interest in the design achievements and burial places of their forefathers and more local and municipal government bodies are becoming aware that they are responsible for a cultural resource which increasing numbers of the young, as well as the very old, believe to be of value.

In Sydney and Melbourne the erection of memorials, mausolea and receptacles of various types is again gathering momentum, particularly as a result of European migration. Unfortunately many trustees and local government officers have forgotten that cemeteries like cities need to be planned and their development guided.

Whatever action is proposed that will affect a cemetery, the first step is to understand in what way, and to what degree, the place is significant. Once this is established plans can be prepared which will retain this significance and may well enhance it. Unless it is understood, any new proposals are likely to be damaging.

Cemetery conservation is a dynamic process. It is likely to involve some preservation, occasional restoration and reconstruction and considerable adaptation, new construction and extension. Modest seasonal maintenance and supervision is necessary to eliminate the expensive periodic repair of cumulative damage. (*Winter 1983*)

Left J. A. Oertel's Rock of Ages, from F. A. Belden, *The crown jewels of art.* Saalfield Pub. Co., 1905

Rouse Hill House

James Broadbent
Photographs by Douglass Baglin

James Broadbent is an architectural historian who has written a number of articles on museums and early domestic architecture. Formerly the curator of Elizabeth Bay House, he is at present on the Historic Buildings Committee of the National Trust (N.S.W.)

Left Rouse Hill House in its tranquil setting.
Above Detail of façade.
Above right The sitting room.

According to family tradition, Rouse Hill House was built between 1813 and 1818, although the grant of the 450 acres of land on which it was built, near Vinegar Hill between Parramatta and Windsor, was not ratified until October 1816. Richard Rouse had begun to develop the property by November 1813 when payments were recorded in the Sydney Gazette for the 'burning of 34 acres of timber, weatherboarding hut' and '60 rods of fencing' at 'Rouse Hill Farm' and building work was certainly being carried out in August 1815 when shingle nails were ordered for the 'House at Rouse Hill'. The house appears to have been nearing completion in August 1817 when a plasterer, William Farrel, submitted his 'Meshurement of Plasterin of the House at Rouse Hill' which lists all the rooms of the present house, and in 1818 Rouse's accounts list payments for four thousand shingle nails, for glazing and for stonework.

Richard Rouse was in a position to build substantially and to build well. A native of Oxfordshire, he had arrived with his wife and two small children in the colony in December 1801 as a free settler. He soon established himself in the Hawkesbury district with a grant of 100 acres at North Richmond. In July 1805 he was appointed Superintendent of Public Works at Parramatta. He was dismissed from this position in 1808 owing to his support of Governor Bligh and he turned his attention again to farming. Governor Macquarie reinstated him in January 1810 and he was appointed auctioneer at Parramatta in October of the same year. He superintended the construction of many buildings, toll-houses and turn-pikes in the vicinity of Parramatta, Windsor and Liverpool. Perhaps most importantly for the appearance of Rouse Hill House he also supervised the renovation of Government House, Parramatta, in 1815–16.

Rouse Hill House is a two-storeyed house of a double range or double pile form — that is two rooms deep — rather than being of a single range or one room deep with a skilling or lean-to behind. It is the only house of this form and scale to have survived from the middle years of Macquarie's governorship. Although the form, first used in the colony at the beginning of the century, was later to become a common arrangement, its use was unusual before 1820 and it was seldom handled with confidence. Indeed, the design of the back range of Rouse Hill House suggests that it was an extension to the original design.

The building works documented prior to 1816 may not refer to the present house, or, alternatively, the long construction period may be explained by Rouse doubling the size of his house during construction, possibly after the recognition of his grant — and after his 'Bilding and Addishon' to Government House. These renovations consisted of enlarging the single range of Governor Hunter's house, 'making it a double house' by adding a series of rooms and a stair hall behind the existing range, and adding wings. The disparity between the front and back rooms at Rouse Hill House suggest that it also was enlarged into a double pile form, possibly during construction.

In plan this disparity is most evident. The principal ground floor rooms, the dining and drawing rooms are in the back range and as well as being larger than the front rooms they are of a different proportion. The back hall, containing the generous staircase, is grander than the front and stone paved; the front hall, smaller and narrower in proportion, has a boarded floor. The internal and (false) external doors into the dining and drawing rooms and the windows of the bedrooms above are awkwardly positioned hard into the corners of the rooms against the back wall of the front range without the customary rib of stone or brickwork against which to fit them. This odd detail may have been a clever expedient if, as postulated, the rear walls were keyed into an existing, completed or partially completed front range for it would have required only half the keying otherwise necessary and it eliminated entirely the difficult keying in of ribs. It was also necessary to position these openings thus in order to maintain a regular window rhythm on the side elevations.

That the back range may have been a later development is also suggested by the construction of the wall dividing the front and back rooms. The thinner wall of the first floor appears to be built flush with the back (i.e. drawing and dining room) side of the ground floor wall, as if for an external wall, rather than being centred on it, and the backs of the fireplaces in the front range protrude into the rooms in the back range on both floors. A change in plan is also suggested by the duplication of fireplaces and flues rather than constructing them back to back.

In its detail the house is consistent externally and shows no sign of two builds but internally the architraves of the front range upstairs as well as down are more elaborate than those in the back range, having blocks at the corners and capped by a narrow shelf.

Finally, the 1818 accounts for stonework record payments of 11 Pounds for two hundred and twenty feet of stone flagging and 6 Pounds for sixteen stone window sills. The flagging was almost certainly the stair hall flagging (which measures two hundred and eighteen square feet) and the sills may also have been for the second range although this has fifteen, not sixteen, windows. (The front range has seventeen). These accounts support the theory that after 1816 Rouse enlarged his existing, incomplete, unplastered and presumably not yet roofed single pile house into the more ambitious double pile house so remarkably preserved today.

Rouse continued to live in Parramatta, and there is another family tradition that originally he intended the house for an inn, although it was never used for this purpose. In 1822 Rouse referred to his 'Country Seat at Rouse Hill' and by 1828 he had retired there to spend the rest of his life as a landed proprietor.

Richard Rouse was also responsible, as Superintendent of Public Works, for 'Making Farneture' for the Government Houses at Sydney and Parramatta and the house still contains several pieces of furniture reputedly made by him which he willed to his son, Edwin: '1 Desk and Bookcase, 1 Dining Table, 2 arm and 6 other chairs', a large 'bed stead complete' and a 'Ward Robe', as well as other important pieces purchased — notably the long case clock in the hall made by Alexander Dick.

Rouse lived at Rouse Hill House until the death of his wife in 1849. He increased his landholdings from an estimated 10,000 acres in 1828 to 230,000 acres by the time of his death in 1852.

As Marjory Lenehan has written in the Australian Dictionary of Biography, 'Rouse was the type of pioneer that the colony needed, a devoted family man, a loyal member of the Church of England, a hard-working and honest public servant and a very affluent grazier'. He was vastly prosperous but, it appears, without the cultural and social aspirations of many of his contemporaries, the parvenus who constituted the colony's elite, the Pure Merinos. The Rouses, disinterested in, and probably, with their emanicipist connections, unacceptable to that flashy soi-disant establishment, lived plainly, solidly and prosperously in their plain and solid house and survived the disastrous depression of the early 1840s which saw the crash of most of the high-flying Pure Merinos and the dispersal of their estates, their rosewood furniture, plate and equipages by the bankruptcy courts, or their refuge under the Insolvency Act. Richard Rouse's continued prosperity however ensured the future of his immediate heirs and it was left to the second and third generations to establish the family socially and to embellish the square Georgian house. The house was not without refinements of detailing and furnishing, for example in the finely moulded joinery and the Dick clock and silver, but it was inferior to the ideals of mid and later Victorian taste and comfort.

Richard Rouse's son, Edwin, inherited the property and it was he who added the verandah in the late 1850s, but it was after his death in 1862 that most of the alterations to the house were effected — by his widow, Hannah, and by his son, Edwin Stephen Rouse (who was a boy of thirteen when he inherited) and by Edwin Stephen's wife, Eliza, whom he married in 1874. The kitchen offices were rebuilt as a large two-storeyed brick wing, the roof of the old house was slated, marble chimney pieces installed in the main rooms which were refurnished and redecorated, the garden extended into the old orchard area and given a decorative latticed summerhouse above the old well and new stables built.

It is a remarkable feature of Rouse Hill — of the house, its decoration and the garden — that subtly, probably unconsciously, each generation has respected the contributions of the former. With social prominence there was no grand rebuilding of the house, no recasting of the plain stone front in fashionable, decorative Victorian stucco or refinishing the rooms with elaborate plaster cornices and florid ceiling roses — just a new entrance doorcase, replacing the old semi-circular fanlight, modern marble chimney pieces and grates and an overlay of graining and marbling, wallpaper dados and friezes and harmoniously toned painted panelling and stencilling over the early colonial fabric. The result is a series of interiors not just of historical importance, but of particular charm.

Australia in Trust

The refurnishing between about 1860 and 1890 and the redecoration, mainly it appears of about 1880, shows a fashionable, affluent, but moderate taste. Nothing is overstated, nothing opulent and although the rooms do not show a family of any particular cultural awareness or attainment, neither are they vulgar. These intimate, comfortable interiors are the most complete examples of mid and late nineteenth century prosperous middle class furnishing and decoration surviving in New South Wales.

A more adventurous and informed taste than is evident in the interiors of the house is shown in the rebuilding of the stables, a dramatic design of spreading roofs and buttressed brickwork for which Edwin Stephen Rouse, in 1876, commissioned the original, if cranky, architect, John Horbury Hunt.

Edwin Stephen's daughter Nina inherited the estate in 1931 on her father's death and lived there with her husband, George Terry, but dwindling family fortunes gradually, almost literally, closed the shutters on Rouse Hill. Mrs. Terry cherished the house, which slowly went to sleep around her. Apart from a new curtain here, and electrical wiring there, which, by their contrast, made the rooms more poignant still, the house appears undisturbed since the First War. Mrs. Terry added nothing and replaced little, leaving the house on her death in an extraordinary balance of preservation and decay. The steamer chair still sits on the verandah as it did in a photograph of 1859, but the olives, planted as garden hedges, untrimmed, have grown into trees, hiding the house from the Windsor Road. But the road is now a busy highway, and the olives are threatened by its widening.

Nina Terry died, aged ninety-four, in 1968, leaving the house to her five sons and by the mid 1970s the property had devolved to her two sons, Gerald and Roderick Terry, both of whom were living at Rouse Hill. From 1976 the sad modern history of the property is recorded in a report prepared in 1982 by the New South Wales Ombudsman.

Left The summerhouse.
Top The drawing room.
Above The dining room.
Right Aerial view of the property.

In June 1977, Roderick Terry's daughter, Mrs. Miriam Hamilton, with her husband, purchased her father's interest in the estate, for which she had obtained a formal option to purchase in April of the previous year. By this time Mrs. Hamilton and her husband were living at Rouse Hill, in the service wing, with her father who was ill. Gerald Terry and his wife continued to live in the main part of the house into which they had moved on Nina Terry's death. It is unfortunate that, quoting the Ombudsman, 'relationships between the two groups were very strained'.

In March 1978, Mr. Paul Landa, the then Minister for Planning and Environment, gave notification of the resumption of the land on which the house was erected. Thus the government compulsorily acquired Rouse Hill House and terminated a family ownership older than any other in New South Wales. Unknown to Roderick Terry or the Hamiltons Gerald Terry had, in January 1977, more than a year before resumption and after the Hamiltons had obtained the option to purchase the father's interest, obtained the following assurance from the Chief Administrative Officer of the Planning and Environment Commission: 'I am able now to give you an assurance that you and your wife would be allowed to continue, during your lifetimes, to live at Rouse Hill House, if the building was purchased or resumed by the Commission, and that no other person would be allowed to live in the house without your consent, other than your brother Roderick, under the arrangements which have existed for the past 7 years.'

The Ombudsman's report states: 'The problem about (this) decision was that it represented the wishes of one party only, namely Mr. G. Terry, and his children. The other party, namely Mr. R. Terry and the Hamiltons, were at that stage entirely unaware of the arrangements which had been made.' Roderick Terry and his daughter were neither consulted about, nor notified of, this decision.

The Ombudsman continues: 'Subsequently, in March 1977, Mr. and Mrs. Hamilton contracted to purchase Mr. R. Terry's interest and in May moved into the part of Rouse Hill occupied by Mr. R. Terry and helped to look after the latter. With the knowledge of the Commission they continued to live there after the resumption took place on 1st March 1978. Mr. Roderick Terry died on the 1st May 1980. On the 24th June 1980, the Commission made a decision that Mr. and Mrs. Hamilton be asked to terminate their occupancy of Rouse Hill House as quickly as possible. This decision was communicated to Mr. and Mrs. Hamilton by letter of 7th October 1980. More recently, upon their refusal to vacate a notice to quit has been issued . . .'.

'The Department (of Planning and Environment, successor to the Planning and Environment Commission) contends that the Commission had in mind a humane and compassionate arrangement to allow the two elderly brothers to remain in the premises, notwithstanding the resumption and that the arrangement enabled the Department to acquire a half interest in the furniture' — for Gerald Terry sold his share of the contents to the government at the time of resumption.

The Hamiltons, however, still own the half share of the contents which are essential to Rouse Hill House's continued importance as part of this country's heritage. They have been denied normal access to these belongings for the last four years and now, despite this ownership of half the contents, despite a recorded statement of February 1978 by Mr. John Whitehouse, the Special Advisor to the then Minister, Mr. Landa, that 'The Minister and the Commission has no desire to dispossess the Hamiltons from their occupancy of the cottage', and despite the finding of the Ombudsman's report that 'the Commission's conduct . . . was unreasonable, and, thus, wrong conduct within Section 5 (2) of the Ombudsman Act', the Department of Planning and Environment is proceeding with legal action to evict the Hamiltons. The present Minister, Mr. Bedford, is not prepared to accept the Ombudsman's finding of wrong conduct and his recommendation to give both parties joint tenancy of the house 'until the deaths of Mr. G. Terry and his wife, or their early vacation of the premises.'

But if the Hamiltons are evicted from Rouse Hill House, what will become of the irreplaceable collection it contains? It would be tragic for the contents to be divided; it would destroy much of that cultural and heritage significance for which, presumably, the place was resumed. Or will the Hamiltons be forced to vacate the house and be expected to leave their share of the contents — half of everything in the house — *in situ* for the life of the uncle who has sold his share to the government, and for the government to use, a result which appears completely unreasonable and immoral. This then is the sad modern history of family disagreement and governmental bungling of Rouse Hill House.

Meanwhile, since resumption only the most basic repairs have been undertaken to the fabric, and these after considerable delays and there is, at the time of writing, still no adequate regular maintenance programme for either the house, the contents, or the outbuildings. It is questionable if the resumption of the property was not precipitate, if the historic family ownership should not and could not have been preserved. The house may have fared better under family ownership then it has during the last four years of government control. The government does not appear to have understood the important and fragile nature of the house and its furnishings, even suggesting that all the contents should be removed to storage while a 'total restoration' is undertaken. Such an upheaval would almost certainly destroy the ambience of one hundred and sixty years of family life.

Conservation will be complex and must be carried out subtly and carefully, but this is a property where no expensive surveys and reports are required to determine its restoration. Rouse Hill is all there mouldering away, complete, but the longer the problems which began with family feuding continue under governmental ownership and specialist conservation work and just plain routine 'housekeeping' are delayed through bureaucracy or indecision or ignorance, the more extensive will the task become and the more will be lost of that unique atmosphere which pervades the tightly shuttered house.

Postscript:

On Wednesday 15 March 1983 Mr. Justice Lee of the Supreme Court delivered his judgments on the actions brought before him.

The Minister for Planning and Environment had taken the Hamiltons to court in order to recover possession of the land and premises; the Hamiltons had cross-summonsed to have their interest in the title returned to them. The Court found the Department of Planning and Environment as being the owner of the premises and entitled to possession. The Hamiltons were granted three months to put their affairs in order and the Department was granted leave to issue a warrant of possession at any time after the expiration of three months.

The Hamiltons also claimed that if, and only if, they failed to have their interest in the title returned to them, or that they were not granted leave to stay at Rouse Hill, then the furniture and contents should be divided.

The judge ordered that the furniture and contents be equally divided in accordance with a certificate of the Master of the Court.

Therefore it appears that, despite the continuing efforts of the Hamiltons, this unique collection will be broken up. One of the principal reasons for resuming the property was that it should be preserved together with its contents. Through its actions, the Government of New South Wales has effectively destroyed what it set out to protect. (*Winter 1982*)

Sydney's first skyscrapers

Len Fox

Len Fox is a former editor of the Miners' paper Common Cause, *and has written a number of books including* Old Sydney Windmills *(1978).*

It seemed to be the best news for many years — a windmill, yes a full-size, working windmill in Sydney Town, a replica of one of Sydney's first skyscrapers of which now not a stone is left in place, but which once dominated nearly every high piece of ground and gave the whole city its character.

The news came early last year, and one could only say 'Too good to be true.' As in a sense it was, for the windmill is not in Sydney itself, but in the rebuilt model of the early city known as Old Sydney Town situated a few hours journey to the north. However, even that is wonderfully good news, for it ends a disgracefully long period where the citizens of Australia's largest city seemed to have turned their backs almost completely on one of the most fascinating architectural and engineering blossomings of its first foundations.

Tasmania has recently claimed as its 'leading tourist attraction' the Penny Royal complex in Launceston featuring a grain-grinding windmill with a watermill thrown in for good measure, and the other Australian States have reminders in one way or another of their ancient mills, but the respectable citizens of Sydney until recently had in one respect seemed determined to maintain the reputation for philistinism which was bestowed on them by J. F. Archibald.

In 1823 William Charles Wentworth had written of 'The lofty windmills, that with outspread sail/Thick line the hills, and court the rising gale' as being among the beauties of Sydney, and as late as 1836 Charles

Above The extent to which early Sydney was dominated by windmills is illustrated by this watercolour (c. 1821), probably by Major James Taylor, of the Military Windmill looking south from Flagstaff Hill. (Photo: Mitchell Library)

Right Major James Taylor. The town of Sydney, 1823. (Photo: National Library of Australia)

Far right Four windmills on Sydney's eastern ridge. Engraving by convict W. Preston from a drawing in 1819 by Captain Wallis from Dawes Battery. (Photo: Mitchell Library)

Darwin had written that as his ship sailed up the Harbour, 'in the distance stone houses, 2 and 3 storeys high and windmills standing on the edge of a bank, pointed out to us the neighbourhood of the capital of Australia...' And for many new settlers coming from the Old World, the first glimpse of Sydney was, as their ship entered the Harbour, the tip of a sail of one of the windmills that stood on the hills around the town.

But by the end of the century, Sydney's mills had been neglected, destroyed, and so thoroughly forgotten that a journalist could write that 'the Sydney windmills are almost as mysterious as the Round Towers of Ireland...' ('The Rambler' in Sydney *Truth,* 20 April 1902). But the very fact that such a comment could be made was sign that Sydney was experiencing a wave of interest in its past. This was to express itself, among other ways, in the foundation of the Royal Australian Historical Society in 1901, and in a paper read to the Society the following year by its president Norman Selfe, which gave the first systematic survey of the nineteen former Sydney windmills.

If ever anyone was worthy of the title of notable Australian it was surely Selfe. Coming to Sydney from England in 1855 as a youth he had that rare breadth of mind which enables a person to become distinguished in the arts as well as the sciences. His career as an engineer and designer and builder of ships, gasometers, workshops, wharves and railways would have been enough for most men, but in addition he became a community leader in technical education and in historical research.

Selfe's paper showed such fascinating glimpses of Sydney windmill history that it seemed almost incredible seventy years later, when I began to take an interest in the subject through the accident of finding myself living in a former windmill area, that no historian had been interested enough to enlarge Selfe's paper into a book. Almost the only detailed research had been that of a writer on the theatre, Eric Irvin[1] through the accident that one of Sydney's windmills, believe it or not, was built on top of one of our first theatres!

One or two other historians, including C. H. Bertie, had also written of the mills, but their work was little known. And yet, as Irvin wrote in his book:

... Levey's mill was inescapable. It towered above the squat Sydney of its day as Macquarie's lighthouse at South Head towered above the 'boxes' at its base. It could be seen from every town vantage point, and from its top

could be seen the ranges of the Blue Mountains. Here was a gargantuan structure whose like could be found in no other British colony. Few of the world's major cities of the time had buildings as high as the total height of Levey's warehouse and mill... The warehouse with the mill on top was in fact some 90 feet high, and possibly higher... The mill itself was a building with six floors and a top portion containing the sail shaft and machinery.[2]

This ambitious structure was the dream-child of a young Jewish businessman, Barnett Levey, who came out to Australia under the sponsorship of his better-known brother Solomon. In Sydney's George Street, at the site where Dymocks Book Arcade now stands, Barnett set out to build a Royal Hotel to include one of Australia's first theatres, with a warehouse at the back, on the roof of which he built his windmill. He was a wild dreamer who ran into stern opposition from the authorities of the day. Many of his dreams failed, and apparently his windmill did not work properly. He himself died young in 1837 in many ways a failure, and yet today he is remembered as 'Australia's first free Jewish male migrant and the founder of the Australian theatre, (who) gave Sydney more glamour, drama, and pathos in his daily life than he ever did on the stage.'[3]

Levey's windmill must surely have been one of the strangest in the world. Yet in 1970 few Australians had heard of it, and a diligent search through the archives gave at first the incredible result that although this had been such a strange and remarkable building, there was apparently not a single drawing, engraving, painting or sketch of it in existence!

Selfe had listed numerous drawings of Sydney's other windmills, but the only graphic representation he could find of the Royal Hotel mill was in the panorama *View of the Town of Sydney* shown by Burford in London in the late 1820s, and even this did not show the completed structure, merely the windmill in the course of erection.

I found it impossible to believe that there was no picture anywhere of such an unusual building, and my disbelief was heightened when I met a University student who assured me he had seen a drawing of the windmill somewhere, though he had forgotten where. So I saddled my faithful Rosinante and rode again through the archives, swearing that I would find windmills though other people could only find imaginary monsters — or was it the other way about?

And a cry of joy rang through the Mitchell Library when at last my ancient eyes alighted on an undated newscutting showing a picture of

Right One of many watercolours by Samuel Elyard of the Waverley mill. (Photo: Dixson Library)

Below An old mill at Parramatta, from *Aust. Picture Pleasure Book*, 1857. (Photo: National Library of Australia)

'Sydney in 1829' dominated by a tall windmill on top of a building in the centre of the city which could only be the Royal Hotel mill. But the newscutting gave no source for the picture shown; it was obviously a reproduction made about 1890 or 1900 of a much earlier drawing or engraving, but by whom? And where could I find the original?

Perseverance was finally rewarded with the discovery of a set of engravings by John Carmichael published in 1829 including the 'mystery' picture and also several other views of Sydney showing the windmill on top of the Royal Hotel. This is surely one of the most interesting sets in existence of pictures of early Sydney — here are beautiful engravings of streets and buildings, soldiers and merchants, convicts, women with parasols, Greenway's Gothic tollhouse that once stood in what is now Railway Square (then a picturesque place of grass-trees, Aboriginals roasting a speared kangaroo, and a coach merrily setting out for Parramatta) — and, from many different angles, the amazing windmill which almost no other artists thought worth recording, but which is here in all its temporary glory.

And yet till recently these historic engravings and their creator were practically unknown. Fortunately, Cedric Flower in his book *The Antipodes Observed* (Macmillan, 1975) has given Carmichael the recognition that is due to him. And even here it had to be written that 'despite thirty-two years spent in New South Wales as an artist-engraver, we know so little about John Carmichael that without the existence of the *Select Views* he might scarcely have existed at all. He is completely ignored in the art histories and national biographies and not much noted by his own contemporaries. . . .'.

'Might scarcely have existed at all. . . .' That could have been said in 1970 not only of John Carmichael, but also of Barnett Levey and of the Royal Hotel Windmill. Today they have at least regained some life and reality. In fact, evidence has been discovered, thanks to research by Sybil Jack of Sydney University's History Department, that after the Royal Hotel windmill had failed, it was probably purchased by Thomas Barker and rebuilt as one of his two windmills at a spot which would now be in the heart of Kings Cross, near where Roslyn Street runs off Darlinghurst Road. And in 1916 a correspondent wrote to J. M. Forde, who as 'Old Chum' conducted an 'Old Sydney' column in the Sydney *Truth*, to say that he could remember building two houses from the stones of one of Barker's mills. These two houses can be identified as still standing today at 42 and 44 Kellett Street; they were due for demolition in the late 1960s but were saved by a Builders' Labourers' green ban and the bankruptcy of a big development company. So if the ghost of Barnett Levey ever walks the streets of Sydney, he might still find a stone or two from his mill. (Anyone interested can still see today the irregularity of the stones in the wall along Kellett Way indicating that they have been used previously in a different building.)

The new windmill at Old Sydney Town mentioned at the beginning of this article is based on a mill much older than Barnett Levey's — the Military Windmill which stood near where the towering Royal Insurance building stands at the southern approaches to the Harbour Bridge. 'We are also erecting upon the high ground over Sydney a substantial and well-built windmill, with a strong tower that will last for 200 years,' wrote Governor Hunter of this mill on 12 November 1796. Within three years it was blown down while still incomplete, by a violent storm. It was rebuilt by 1802, but by the time of Commissioner Bigge's Report in 1823 windmills had begun to give way to steam power. By 1902 the Military Windmill was remembered only by a few historians like Selfe as a landmark that had stood 'just south of the Grosvenor Hotel.'

By the time I began my research in the early 1970s I couldn't for some time discover anyone or anything in Sydney to tell me anything about the Grosvenor Hotel or where it had stood. How quickly a city can change! How swiftly important buildings can be almost completely forgotten!

The Grosvenor, as I finally discovered in the Mitchell Library archives, had been no tinpot little pub. In 1902, when Selfe was delivering his paper on Old Sydney Windmills, the Grosvenor was one of the city's largest and most luxurious hotels, boasting in its advertising matter of its 'pure air' (yes, how quickly a city can change!) and a Gentlemen's Smoking Room which was perhaps 'without parallel in the world.' Now, not a stone is left. And hardly a memory.

But at least it is good to look at early pictures of Sydney, and see how the Military Windmill and the nearby Wooden Mill were the skyscrapers along Sydney's Western Ridge, just as Boston's and Palmer's mills, and the smaller wooden Kable's mill, dominated the Eastern Ridge, standing where three bronze statues now stand in or near the top part of the Botanic Gardens (where another near-forgotten Sydney structure, the glass Exhibition Palace, once stood). The Mitchell Library has some attractive pictures of these mills, including a fine engraving by convict W. Preston from an 1819 drawing by Captain Wallis. This engraving appears to show four mills on the Eastern Ridge, but Selfe has pointed out that one of them is a mill known as Clarkson's Mill at Darlinghurst seen in the distance.

Clarkson's Mill stood near the Gaol (East Sydney College of Technical and Further Education), probably in the area bounded by Liverpool, Burton, Darley and Forbes Streets, an area with many sandstone walls probably built from former windmill stones. It was one of a number of windmills along the ridge which extends from Potts Point out to Waverley — all gone now, and all almost completely forgotten, except for an odd memory here and there, like Mill Hill Road in Waverley with the Church of St. Barnabas carrying the name of 'Mill Hill Parish' and with a framed drawing of the old Waverley windmill on the vestry wall.

At the other end of the former 'Sydney windmill belt' is Miller's Point, probably named after a legendary Jack the Miller who, so it is said, could once have got the whole area for nothing if he had offered to put a fence across the neck, and who, according to Norman Selfe, was in reality John Leighton who died in 1826 and whose grave at Botany is not far from that of Barnett Levey. Here too is Windmill Street — and yet the whole area has been changed so much that the names don't mean much.

It has been good in recent years to see people organising to defend the buildings of the 1890s or the 1920s and 1930s, but perhaps the time has come also, when so much of Australia is being turned into forests of block-like skyscrapers, to encourage a fresh wave of interest in our earliest skyscrapers, the windmills whose clean and natural source of power had a lesson for future generations.

E. W. Marriott's recent book *Thomas West of Barcom Glen* (Barcom Press 1982) shows us how closely the early history of Sydney is tied up with the story of its mills (in this case a water mill). We could do with many such books. And on the artistic side, there is surely room for books showing the paintings and sketches of artists like Samuel Elyard and John W. Hardwick whose work captured so much of the atmosphere of Sydney of last century.

And in the meantime we can be thankful for Old Sydney Town. To have built a full-size working windmill faithfully incorporating all the features of the original including the wooden cog-wheels, at the same time keeping within today's safety standards by incorporating hidden modern safety devices, was an extremely difficult and costly architectural and engineering problem, and the project has won praise from such world authorities on windmill reconstruction as British expert Ernest Hole. If Norman Selfe is to be honoured for his work on Sydney windmills, so also should we remember the name of Fred Whitburn, the retired marine engineer who in the last years of his life, though dying from heart disease, continued as Operations Manager to supervise this construction as his great dream.

It is an inspiring example of what can be done to preserve the past even when the past no longer exists. The Military Windmill, which was to have lasted 200 years but blew down in three, and lasted only about thirty years in its second life, is now enjoying a third innings in which it is determined to prove that after all it can last, as Governor Hunter predicted, till 1996 — and perhaps much longer! (*Summer 1983*)

**Mt Gilead windmill near Campbelltown.
(Photo: N.S.W. Govt. Printing Office)**

Twentieth century buildings

Peter Spearritt

Peter Spearritt teaches urban politics and urban history at Macquarie University. He is the author of Sydney Since the Twenties *(1978) and* Sydney Harbour Bridge: A Life *(1982) and co-author of* Australian Popular Culture *(1979),* Sydney: A Social and Political Atlas *(1981) and* The Open Air Museum: The Cultural Landscape of NSW *(1980).*

Twentieth century buildings are common. Too common for their own good. If they were fewer in number, historians and architects would take them more seriously. As it is, we are surrounded by them, bored by them and often oppressed by them. Little wonder that when they are demolished nobody gives a damn.

Ironically, nobody gave a damn when most of Macquarie Street's elegant stone terraces were demolished in the 1920s and 1930s to make way for buildings like the Astor Flats and BMA House. The lesson we can learn from such an observation is that people rarely appreciate their immediate past history. Why did it take over one hundred years for anyone to get upset about the demolition of 1850s stone terraces? Partly because until they started to be demolished they were pretty common. But what a difference a few years of demolition can make.

Historians and architects are only now beginning to re-examine the assumptions under which particular buildings and groups of buildings are thought worthy of restoration and preservation. When J. M. Freeland's *Architecture in Australia: A History* first appeared in 1968, few commented on the fact that the book had hardly anything to say about 20th century buildings. The conventional wisdom of the time was that the great days of innovative Australian architecture ended somewhere around 1880. With the notable exception of a few inspired individuals like Walter Burley Griffin and Harry Seidler, this country had produced little of significance since the halcyon days of Francis Greenway and James Barnet. Such attitudes were also shared by the National Trusts in each state, which were then bestirring themselves mightily in the fight to save the best examples of our nineteenth century heritage. Crudely put, if you wanted a building saved, your best bet was pre-1900 and in stone. If it was made of wood, you didn't have much hope unless it was pre-1850. If it was made of fibro and post-1950 the less said the better.

The greatest obstacle confronting the would-be preserver of 20th century Australian buildings is that many people — including many conservationists — remain unconvinced of the claims of this century's buildings for preservation. One reason is that very little has been written about 20th century buildings, apart from day-to-day comment in the architectural and building press. A number of country towns and metropolitan suburbs now boast adequate histories, though these understandably are usually more concerned with the human than the built environment. Particular types of buildings have attracted attention, especially if the type is aesthetically pleasing or is thought to reflect the spirit of the times. Thus Ross Thorne's *Picture Palace Architecture in Australia* (1976) and his equally evocative but meticulously documented *Cinemas of Australia via USA* (1981) have drawn our attention to the central role, both in terms of architecture and entertainment, that the cinema played in our society. And we can lament with Thorne that so many of the great

Twentieth century buildings

Left While Murgon's Busy Bee Cafe has gone for good, Innisfail's Blue Bird Cafe, dating from the mid-1930s, still retains much of its original character.

Above This late 1920s stone cinema and adjoining R.S.L. Hall in the working class Adelaide suburb of Semaphore have remained remarkably intact.

movie houses of yore have succumbed to the siren song of skating rinks, furniture warehouses and restaurants. At least some striking examples of cinemas remain, though we could certainly do with more of them.

A number of other building types have also attracted attention, especially if they were the work of big-name architects. Thus the work of Bruce Dellit (including the War Memorial, Hyde Park) and Emil Sodersten (including the War Memorial, Canberra) has recently been celebrated by contributors to *Architects of Australia*. Even in Donald Johnson's recently published and eminently useful *Australian Architecture 1901–1951: Sources of Modernism* most attention focuses on the big name architects with the vernacular getting short shrift indeed. We desperately need a study like Robinson and Bletter's *Skyscraper Style: Art Deco New York* (1975) where William Van Alen's Chrysler Building jostles side by side with apparently nondescript office blocks.

But what of those 20th century buildings that have no one to champion them — service stations, cafes, fruit shops and butcheries, to name but a few? Such buildings are the stuff of everyday life, but being modest structures subject to the whims of taste and the market place, they are easily demolished or renovated. It can happen so quickly that no-one has time to complain.

Top This turn of the century church and rectory at Narooma on the N.S.W. south coast can be either delightful or nondescript, depending on your point of view. With declining congregations, these buildings are now becoming just as subject to redevelopment pressures as many other buildings.

Above When this Cairns bowling club site was advertised for sale in May 1980, not even local preservation groups bothered to give the c. 1910 brick, wood and galvanised iron structure a second thought. The pressures for redevelopment of the site are strong.

Right This elegant war memorial at Lithgow is much more likely to survive than the late 1920s cinema opposite, which will most likely succumb to a changed use.

It is virtually impossible to walk down the main street of a sizeable country town or suburb in Australia without seeing at least one shop undergoing 'renovation'. Chances are that this shop will have been built in the twentieth century — most Australian shops have been. Katoomba, one hundred kilometres west of Sydney, and Murgon, 250 kilometres north-west of Brisbane, have very little in common, save that they both once boasted elegant interwar cafes. One is a mountain resort and guest house capital, the other the centre of cattle and farming. Katoomba's magnificent art nouveau 'Paragon Cafe', c. 1916, is justly famed. Little has changed since the last wave of renovation in the 1920s. Less well-known was the 'A B Cafe', opposite the railway station. Possibly the best example of a mid-'thirties cafe in Australia, replete with American style chrome soda fountains, it was completely gutted and remodelled in 'laminex and chipboard moderne' in the late 1970s. At least it is still a cafe. Murgon's 'Busy Bee' Cafe has suffered an even worse fate. In January 1979 it had changed little since the early 1920s. Its art nouveau leadlight cabinets were still in place, as was its Wunderlich pressed metal ceiling, a relic from an earlier era. A year later both the facade and the interior were completely gutted for a furniture shop. According to local businessmen the fittings have gone to antique dealers far and wide, while

the facade went to the tip. There is not much of a market for antique leadlight in Murgon.

Why, many people ask, should one be upset by this process of renovation. After all, they argue, the market place is paramount, and businesses must keep an up-to-date image. But both these cafes were viable propositions and their proprietors could have turned their unusual decor to good advantage. But the pressure to follow next-door's new aluminium plate-glass windows is strong.

The worst architectural vandals in Australia are not, however, the small-time operators, but the big-time chains. Coles and Woolworths have between them ruined many a main street in many a country town. The magnificent Coles Stores of the interwar years, with their black tiles, blended well into country shopping centres. But the new wave of K-Marts, Coles New World Supermarkets and Big W's are rarely designed with any feel for their setting. Blank brick walls often front onto main streets, replacing more intimate and often better-built shops of earlier decades. The banks, too, have been responsible for some ghastly renovations, often carried out in country towns where business is so slow that there can't possibly be any pressures to update. The CBC and the Wales are finally starting to realise that their nineteenth century banks are worth preserving, but as for their 20th century structures, they rarely care. The Commonwealth Bank, a product of this century, has a penchant for putting cheap facades over its own once elegant structures. Even the R.S.L., seemingly more immune from commercial pressures, has often 'beautified' its own interwar facades by adding a besser-block feature wall.

Australia's cities have witnessed a number of waves of demolition. In Sydney and Melbourne, nineteenth century terraces were demolished in the interwar years to make way for new industrial plant in the inner areas, while houses were demolished in middle ring suburbs to make way for flats. Most building in the 1950s in all Australian cities consisted of bungalows on virgin acres. It was not until the 1960s that we saw a demolition and rebuilding programme to rival that of the interwar years. In the last two decades the face of the central business district of every large Australian city has been altered irrevocably, with two exceptions, Newcastle and Canberra. Central Newcastle still retains much of the character that it had by 1940: the great bulk of its buildings as of 1982 were built between 1880 and 1940. Central Canberra, on the other hand, is largely a creation of the last twenty years, while most of its most substantial pre-1940 buildings, such as the Sydney and Melbourne arcade buildings, have fortunately survived.

The position in the other cities is not a happy one, and while most public attention has been focused on demolition in the central business districts, probably greater havoc has been wrought in the suburbs through indiscriminate demolition for flat development. Who but old timers can remember the charm of Sydney's Wollstonecraft or Brisbane's New Farm before flat redevelopment took place?

One of the reasons why much of this redevelopment went ahead not only unhindered, but with little outcry from preservation groups, was that the developers were not knocking down 1850s stone terraces or 1880s polychrome brick-work churches, they were knocking down federation style and Edwardian homes, pre-war shops with art nouveau facades and interwar buildings of all types: funeral parlours, garages, textiles factories, and even small blocks of flats.

When large-scale redevelopment for flats, drive-in shopping centres (among the greatest eaters of space) and single-storey petrol stations began in the late 1950s they carried all before them. Local councils who were used to the virgin acres activities of spec builders, owner builders and the Housing Commissions took a long time to realise what was happening. In many cases the councils themselves were dominated by people with a vested interest in local real estate or local commerce, people who were happy to fuel the apparently inexhaustible process of demolition and rebuilding. Ten years later many local residents began to wonder what had happened to their old streetscapes. Even many residents of the new flats, beneficiaries of the central locations which development had provided them with, began to question the redevelopment process.

The National Trusts in each state fought valiantly against much mindless demolition, but without statutory powers they had their greatest successes with buildings thought by the public to be of historical and/or architectural interest. In general, 19th century buildings fared better than 20th century ones. The creation of the Australian Heritage Commission and of Heritage Councils in some States in the mid-'seventies heralded a new era of concern, but unfortunately the national body does not have the statutory powers enjoyed by its N.S.W., South Australian and Victorian counterparts. The Heritage Councils/Committees are now embarked — in conjunction with the National Trusts and architectural institutes — on a programme of much-needed research designed to document key 20th century structures. The recently published *The Heritage of Australia: The illustrated Register of the National Estate* includes many notable 20th century structures. But the Councils and the Trusts are still confronted with two over-riding problems: how does one assess 20th century buildings and how does one justify preserving them?

Beauty, as we are often told, is in the eye of the beholder. When it comes to buildings or cultural landscapes there are so many beholders, with such a variety of backgrounds and vested or altruistic interests, that a concensus is often difficult to find. Moreover many people concerned about preserving Australian urbscapes baulk at 20th century buildings, in part because there are so many of them. How on earth is one to establish criteria about what to preserve, when over three quarters of all extant structures in this country have been built since 1900. It is all very well to dismiss 'preservation criteria' as a bureaucratic plot, but when confronted with such a vast task, such criteria are sorely needed.

Criteria for preserving particular buildings are usually hotly contested, but they can be reduced to five main categories: aesthetic, architectural, technological, historical and community needs. Nor are these categories discrete. They may overlap or one may predominate.

The arguments that rage over the preservation of 19th century buildings become even more uncertain for the following century. More and more people are becoming convinced that on architectural, technological and historical grounds, certain categories of 19th century buildings should be retained. There is much less agreement about the role of community needs in all this. Who should use, live or have access to the buildings? The Rocks, for instance, now provides the odd spectacle of Housing Commission tenants jostling with wealthy and not-so-wealthy tourists for pedestrian space. Bitter arguments are now brewing about who will be able to afford to rent or buy the flats currently being constructed within some of Sydney's early 20th century woolstores. The developers argue that their flats will be available to medium income earners, while critics assert that only the rich will be able to afford the prices. Many others couldn't care less what the price is, they just could not stomach living in a renovated wool store, behind liver-coloured facades they consider both oppressive and outdated. In the case of the woolstore renovation, as distinct from restoration, the use to which the building will be put has changed drastically. While there may be no agreement on aesthetics or community values, many preservationists would argue that there are historical and architectural arguments for keeping the facade. For the developers, retaining the basic structure makes economic sense in an age of high building costs. Preserving the facade will also preserve the historical and architectural 'feel' of this part of Pyrmont, but it will do little for those interested in technology, as the actual mechanics of the woolstore will be virtually obliterated in the renovation.

Technological concerns are only now coming to the fore. For a country that persisted until recently as seeing itself as free of industrial blight, Australia has a remarkable array of industrial architecture and plant, most of it dating from the early 1900s. BHP's plants at Newcastle and Port Kembla still contain sections little changed from the interwar years. Arguments about preservation for purposes of the history of technology in such plants usually fall on deaf ears, especially from the owners of the plant. Given the unassailable ethic of profit-maximisation, perhaps governments should consider granting tax concessions to those functioning industries who are prepared to preserve some plant for public inspection (itself a difficult proposition in many cases).

The arguments for preservation in the case of abandoned plant are much less complex. The recently-produced documentary on the Glen Davis shale oil plant (north-west of Sydney), is a timely reminder of our 20th century industrial heritage. Here is a magnificent industrial relic, Australia's answer to Britain's sacked monasteries. Judging from audience reaction, most Australians are astounded to find that 20th century Australian technology can be so interesting.

Bunnerong Power Station, built by the Sydney City Council in 1929, is an example that falls half-way between the Glen Davis and BHP plants. Thought by 1976 to be obsolete (it had been used on only a handful of occasions in the early 1970s as a back-up power station), the incoming Wran government negotiated to sell its site to a petro-chemical combine. Demolition of the huge structure, probably Australia's largest pre-war industrial building, was a slow operation because of the enormous amount of scrap-metal within it. Unmoved by arguments about its preservation as a technological relic, the Wran government is now thankful that the demolition was such a slow process. What's left of Bunnerong is being revamped with new turbines to provide sorely-needed back-up in any N.S.W. power crisis.

When it comes to preserving 20th century buildings, I argue that we need to save not only such unique structures as the Bunnerong Power Station and Melbourne's Forum Cinema, but representative examples of common buildings that, given urban redevelopment, will not always be so common. Such buildings should be seen *in situ*. There may be little point in preserving a two-storey art nouveau cafe if it is dwarfed by high rise towers. If we are to preserve the commonplace as well as the unique — to give some idea of the everyday built environment in times past — streetscapes are often as important as individual buildings. It is easy to be clever with hindsight about what should have been saved. Preservationists today need to develop foresight. Who knows which of our present-day structures will be a source of fascination to future generations?

(*Winter 1982*)

Mulwala homestead complex

Peter Freeman

Peter Freeman is the author/illustrator of: The Murrumbidgee Irrigation Area Sketchbook, *Rigby, Adelaide, 1978;* The Woolshed: A Riverina Anthology, *Oxford University Press, Melbourne, 1980;* The Homestead: A Riverina Anthology, *Oxford University Press, Melbourne, 1982.*

Beside one of the many lagoons that adjoin the Murray River stands the Mulwala station homestead complex. Mulwala station, situated four kilometres west of the village which gave its name to the station, is roughly equidistant from the Murray River townships of Corowa to the east and Tocumwal to the west. Corowa, Tocumwal and Mulwala townships were all settlements that had their origins in neighbouring pastoral runs, and the river east and west of Mulwala is dotted with the homestead buildings of once great runs. Across the river from the Mulwala (originally Mulwalla) run, was Yarrowey (later Yarrawonga) a run which was taken up in 1840 for Elizabeth Hume by her brother-in-law Hamilton. Up the river to the east was Collandina (later Collendina) run which was taken up by Brown; and

down the river to the west was Boomoonomin (later Boomanoomana) run taken up by Jeffries; Burrogo (later Barooga) run taken up by George Hillas; and Tocumwal run taken up as part of the mighty Riverina empire of Benjamin Boyd. All these runs were taken up and claimed by the 1850s, so when Alexander Sloane from Melbourne established the Savernake run in 1862, and the Mulwala run in 1864, they became the younger sisters of the larger and older neighbouring runs.[1]

Alexander Sloane was born in Glasgow on 29 May 1829, the son of James Sloane, a merchant of Glasgow. He was educated at Glasgow High School, and in 1849 he migrated to Australia with his brother William. After a period of employment with Frederick Dalgety's new pastoral company, Alexander Sloane ventured to the Victorian gold-fields where he was unsuccessful. He returned to Melbourne, formed his own pastoral company, and in 1853 while in Great Britain on business, met his future wife Annabelle Helen Gibson at Taunton in Somerset. Sloane returned to Australia by the steamship *Argo* in 1854 and one year later was joined by Annabelle. The couple were married at St. Kilda the following year, and they remained in Melbourne until 1860. After a brief period in the Western District of Victoria (Major Mitchell's 'Australia Felix'), Sloane's attention turned to the pastoral run north of the Murray River. On 28 August 1862 he purchased Savernake stock and station north of Mulwala village from a Mr. Graham.[2]

Sloane began construction of a house at Savernake and superintended the shearing of his newly-acquired stock. By December the new station was in readiness for the arrival of Annabelle and the four Sloane children from Melbourne. The 230 mile journey took the family nine days (with stops at Wallan Wallan, Seymour, Avenee, Longwood, Baddaginnie, Wan-

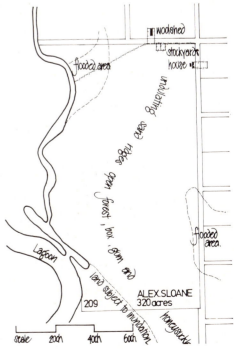

Bottom left Mulwala station homestead, c. 1871.

Bottom 'Plan of a portion of Land No. 209, Parish of Mulwala, County of Denison. Applied for as a condition purchase by Alexander Sloane.' Transmitted to the Surveyor-General 21 October 1869, John Burns, licensed surveyor, Department of Lands, N.S.W.

Below Sketch by Surveyor Thomas Townsend, 19 May 1857, showing the first settlement at Mulwala village — McCrae's Inn, Surveyor-General: Surveyors Sketch Books, Archives Office of N.S.W.

garatta, Wahbunyah, O'Bryan's, and Savernake), and cost £18. The cost for cartage of their household effects was £15 per ton. The next year Annabelle's mother set out for Savernake and had an equally formidable journey. Once at the new home station she quickly became involved in homestead life, as she related in her journal:

There were no vegetables, milk or butter on the Station. On Sundays or when a beast was killed, we had a pudding. The cook made good bread and on baking days we had fried dough. I looked forward to these mornings like a child. How the children survived I do not know. Strong tea without milk, sugar that made it like syrup; I have seen Janet aged 13 months eating a piece of suet pudding at 8 in the morning.

Mrs. Gibson also assisted with the completion and decoration of the Savernake homestead:

The sitting room roof was only half finished. Alexander got the splitters to hurry up the work and they made a good job of it. The sitting room was lined with calico. Annabelle cut wall paper, I pasted and helped Alex put it on. I covered the mantel piece which was made of packing cases, the labels still being visible. The table was made of boards fixed on saplings for legs, cases made a nice sideboard with curtains of American cloth.[3]

On 10 May 1864 Sloane took delivery of Mulwala station to the south of Savernake which he had bought from a Mr. Hillas. (Hillas also owned Barooga run to the west.) Mulwala was then bounded on the east by Teremiah station and on the west by Boomanoomana station. It had a frontage of four and a half miles on the Murray River and extended north thirteen miles to Savernake run. Once again the Sloanes and their now considerable entourage had to move house.

Sloane took over Hillas's homestead, which was near the Mulwala lagoon, adjacent to the village. The first three years were drought years, broken on 7 October 1867 by a devastating flood that carried away the homestead kitchen, chimney, and front part of the main homestead. Again the Sloanes moved, this time to a sand-hill site Sloane selected about a mile north of the earlier homestead. Here the Sloanes commenced the homestead buildings that stand today.[4]

The early 1870s brought the completion of the railway line from Melbourne to Wangaratta and beyond, and with it an influx of selectors eager to take up agricultural land on the old runs. The Sloane family were equally eager to secure their land and both family and friends took up selections at this time. The influx of selectors brought an increase in village population and the need for more buildings and station improvements. In 1883 Sloane began building a sawmill to mill timber for his own property and for the Mulwala village. By 1887 the village had grown considerably, boasting a court house, two hotels, two stores, and Sloane's sawmill.[5]

By the late 1880s Mulwala was at last an established station; most of the brush fences had been replaced by seven-wire fences, and dense forests had been cleared for roads and paddocks by ringing, suckering, and clearing, and the sheep were producing more wool due to Sloane's careful breeding and classing. However, in the 1890s depression struck

Top Ante-room with stag head. The interiors of the original building are all lined with Murray pine 'slab and batten'.

Above The Murray pine lined dining room.

Right The drawing room. The Murray pine boarding has been wallpapered.

and sheep prices plummeted. The sheep that had prospered in the past good seasons were sent to boiling-down works at Corowa where they were sold for only 2s 9d per head. Sloane turned to wheat farming on his sheep run and brought in share farmers to help him. Farming saved Mulwala and allowed Sloane to maintain his station even during the great drought that followed the economic crisis. While banks foreclosed on stations around him Sloane continued at Mulwala.[6]

On 12 October 1907 Sloane died in Melbourne at the Grand (now the Windsor) Hotel. The station he had taken up as a timbered wilderness in 1867 had been transformed into an efficient and well-established pastoral and farming property. The original Mulwala station has since been much divided and diminished but the original homestead block is still owned by a descendant of Alexander Sloane, and the surrounding country and village streets of Mulwala bear testimony to the Sloane name.

Having been flooded out of the first Mulwala homestead in October 1867, Sloane immediately began his new building:

Jimmy Neil was the carpenter, he was employed for many years as a carpenter and a carpenter's shop was built for him at the new homestead. He lived in Mulwala township and walked down every morning [$2\frac{1}{2}$ miles away] ... He started taking down the old house for removal on 4 March 1868, and he and Sutherland finished the new house on 30 May, and on 24 June he started putting up the stables, which had been brought over with the bullocks.[7]

In 1869, the homestead portion survey was prepared by John Burns, a surveyor. The survey showed three distinct homestead buildings crowded on the east by the surveyed allotments of Mulwala village. Three years later a correspondent from the *Town and Country Journal* visited the station. The correspondent's description was illustrated with a fine wood engraving which accords closely with a contemporary photograph of the homestead:

This residence is one of the most pleasant on the Murray. It comprises several buildings, in cottage style, though much more extensive than such buildings usually are, and thoughout they are built of pine. The Murray pine I have already spoken of for its beautiful grain, and when polished it is perhaps unequalled among Australian timbers . . . on looking at the interior of Mr. Sloane's residence, with its polished ceilings and panelled walls, the visitor sees the justice of the appellation 'cedar palace'. There is a good shrubbery and garden, filled with choice plants and flowers before the residence, and a cheerful view from the verandah over a stretch of green fields and meadow lands to the entrance gates may be had. Station life must flow quietly and pleasantly at Mulwalla.[8]

The photograph and engraving also show three adjacent buildings, all with pine slab walls and shingled roofs. The kitchen, the smaller cottage on the south, had a metal chimney and brick floors. All the internal walls were lined with Murray (cypress) pine boarding, finished with calico and wallpaper. The ceilings were lined with calico, for John Sloane recalled that in 1886 a Melbourne carpenter named Thomas Porter 'came to Mulwala and started putting up the boarded ceiling in my father and mother's room at the homestead. It looks as if he put up all the ceilings . . . which were previously of calico.'[9]

By 1899 the old kitchen block had been demolished and a new kitchen was built to the south. New bedrooms and a schoolroom were built where the old kitchen once stood. In that year an English visitor arrived from Melbourne and described the homestead in a letter to her parents:

Half a mile further we reached another white gate which was the entrance to the 'Homestead'. To the right was a long low house with verandah running from end to end. To the left facing it was a beautiful garden, flowers near the house and trees and fruit further off . . . Behind the main building are numbers of other houses and offices finishing up with the men's quarters, woolshed, stables, etc. which I have not yet visited . . . Now sitting as I do on the verandah I see plots of flowers, edged with a wide border of buffalo grass. This grass is coarse and only green in the summer time and becomes brown in winter . . . In the garden are grapes of every kind, green and purple, muscatels, etc. All kinds of vegetables from parsley and carrots to water melons the size of a church hassock, and pumpkins, rock melons, tomatoes, and quinces . . . Now I wonder if you have any picture of Mulwala in your mind's eye![10]

Mulwala homestead continued to grow and evolve. A pisé wing designed by the Corowa architect MacKnight was added in 1927, and a brick addition was built to the south. To the east, between the station homestead and Mulwala township, the Mulwala Explosives Factory was established during the Second World War. By the 1970s, the homestead proper had been vacated and only the modern southern wing was still being used as a dwelling. The homestead complex, still owned by a member of the Sloane family, had become isolated between the Murray lagoons and the Mulwala Explosives Factory, on land difficult of access and inefficient as the headquarters for the farming and pastoral operations of the station.

The threat that now faces the homestead complex is the intention of the Commonwealth Department of Industry and Commerce to acquire the homestead and woolshed portions of the station to allow enlargement of the Explosives Factory. It is to be hoped that satisfactory measures can be ensured for the continuing conservation and management of this once great Murray River station homestead. (*Winter 1982*)

Harrisford, Parramatta

Clive Lucas

Clive Lucas, O.B.E., is a Sydney architect well known for his work in conserving Australia's heritage. Here he writes about the restoration of one of Australia's oldest houses.

Above The front hall retains its 1820s plaster ceiling and fine cedar joinery. The mid-Victorian blocked rusticated wallpaper was reproduced from samples discovered behind later linings fitted in this room.

Above right The main façade is characteristically Georgian and balanced. However a closer inspection reveals that the window and corner piers to the right of the door are wider, indicative of the widening of the house to match the alignment of the school room added in 1832. (Photos: Max Dupain)

'One year's rent on a two-storied house at Parramatta, £75.0.0'[1] was listed as one of the assets of the Rev. William Walker (1800–1855) on 17 July 1830 in his application for a grant of land at Bathurst. 'Mr. Walker', as the Land Commissioner's report stated, 'arrived in the colony as a missionary connected with the Wesleyan Methodists in the year 1821, but he did not long continue in that connection. He has for the last three or four years taught a School in Parramatta and is married to a daughter of the late Mr. R. Hassall.'[2] Walker had, in fact, fallen foul of his church because of the property he had accumulated, largely through his marriage into the Hassall family.

Parramatta was a government town and the land on which Harrisford stands was a leasehold first granted to an emancipist publican, William Carter, in June 1823.[3] The front section of the present building was probably built by Carter about the time he acquired the lease and is probably that shown on Stewart's *Map of the township of Parramatta* made in the same year.

On 14th May 1823 William Walker married Elizabeth Cordelia (1804–1835) the second daughter of missionary Rowland Hassall (1768–1820) whose home was in George Street immediately opposite. Walker was suspended from the Aboriginal mission in 1824 and thereafter became Superintendent of the Female Orphan School at Parramatta but resigned in April 1826 because of mounting criticism, particularly from Archdeacon Thomas Hobbes Scott under whose jurisdiction the Orphan School came. It was then that Walker opened a school of his own and it is quite likely this was conducted at Harrisford, which he rented from William Carter.

In September, 1829[4] Walker purchased the leasehold of the two lots which now form the basis of the property. The back section running down to the river was purchased from Joshua Allott for £15 and the front section facing George Street, with the building on it, from Carter for £60. Soon after Walker mortgaged his new property to Samuel Terry for £500 and in the papers mention is made of the 'dwelling house, standing . . .'.

Although by 1828, he had large pastoral interests on the O'Connell plains near Bathurst, Walker was still listed as living at 'George Street-Parramatta' at the end of 1831,[5] although in July 1830 he had stated that he was removing himself and his family to Bathurst.

The King's School at Parramatta is the oldest continuing school in Australia. One of the tasks of Scott's successor, Archdeacon William Grant Broughton, on his appointment to N.S.W. in 1829 was to establish King's Schools based on 'the system long established and proven in the Public Schools of England'. A King's School was established in Sydney and another at Parramatta. The one at Sydney failed, but the one at Parramatta survived.

In March 1831 Viscount Goderich, Secretary of State for the Colonies, gave approval to Governor Darling to establish the two schools[6] but it was not until the appointment of the Rev. Robert Forrest (1802–1854) as headmaster and following his arrival in January 1832, that the Parramatta school could open its doors.

Harrisford had been leased from Walker for £60 per annum and its first three pupils were admitted on 13 February 1832. Within six weeks there were sixteen, and on 3 April the headmaster ordered desks, forms and other items of school furniture.

By the second term demand for places was such that further accommodation had to be provided and on 9 July the headmaster and the Rev. Samuel Marsden entered into a contract with two Parramatta contractors, James Byrnes and Robert Gooch, to build a 30 × 18 foot school room and other rooms for the sum of £120.10.0.[7]

The school continued to prosper and two adjoining houses had to be rented for bedrooms before the school moved to the new stone building specially built for it, further up the river, to begin the school year of 1836.

From the start King's attracted the sons of pastoral families which has continued to this day. Names like Ryrie, Oxley, Blaxland, Rouse and Cox were amongst those educated in Harrisford. In 1835 there were 124 pupils, 96 of whom were boarders.[8] The accommodation of so many students would suggest that more was added to Walker's house than is mentioned in the contract with Byrnes and Gooch.

The two-storey house acquired by Walker in 1829 would appear to have been the archetypal single pile Georgian brick house, five bays wide, its walls enhanced by freestone quoins and string courses, centred on a fanlit front door all under a simple hipped roof. The plan of the front hall suggests that it was meant to contain a simple single flight stair although no physical evidence of such a stair has been found.

No matter, a school room 30'0" long × 18'0" 'in clear' was built and it seems immediately after this a stair hall and a second storey over the school room were added. At this time it appears the drawing room (i.e. the room to the right of the front door) and the room above were widened to correspond with the alignment of the side wall of the school room and the whole side roofed with a fine queen post trussed roof. The wall construction to the right of the front door is different to the rest of the facade[9] and the corner and window piers are wider. This growth is supported by surviving joinery details and an analysis of mortar and plaster samples taken from the building.[10]

Above The library wall uncovered revealing built-in cedar book shelves with apsidal heads covered in three layers of Victorian wallpapers.

Above right The front hall stripped down at the start of the restoration. (Photos: Clive Lucas)

The additions have simple butt and mitred architraves whereas the original section has architraves which are more elaborate and finely reeded. This refinement is greatly enhanced by the chimney wall of the library, where the breast, enriched by a shallow arch, is balanced by spherical headed niches holding book cases. (The Rev. William Walker, who had a library of 1,100 books, was described by Governor Brisbane as the best educated man in the colony.[11]) Such architectural detail suggests educated taste and it is conceivable that Walker was responsible for fitting out this section of the house during his occupation.

There must have been a detached kitchen and in 1832 mention is made of a coach house and stables but no trace of these has survived. For some time after the completion of the additions the building seems to have been 'L' shaped with a stone flagged yard in the remaining space.

Harrisford continued as a school. In July 1836 Mr. J. Bradley announced that he was moving his academy to the 'spacious and eligible premises formerly occupied by the King's School.'[12] Little is known of Bradley's academy but the next school to lease the property is better known.

In 1841 the Rev. Dr. William Woolls (1814–1893), schoolmaster and botanist, who had taught at King's while it was at Harrisford, set up his own school. This academy, which attracted the sons of influential families, remained at Harrisford until the late 1850s when it was moved to Broughton House.[13]

In 1852 William Walker had secured the freehold of the property and the following year sold it to Michael Eury, master mariner.[14] It did not change hands again until 1890 when Eury's executors sold it to John Harris (1805–1891) of Shanes Park near Penrith, a nephew of Dr. John Harris of Ultimo and Harris Park.

After Dr. Woolls' it was the home of the Linden House Girls' School[15] run by Susan Griffiths (1808–1898), the wife of the colonial artist William Griffiths (d. 1870) who was drawing master at the King's School between 1839 and 1842. Mrs. Griffiths' school occupied various buildings in the town and finished up in a stone house in Macquarie Street which has since been moved to Lancer Barracks.

After Harris acquired the property he named it Harrisford and converted it into a residence for members of his family. As part of this work a two-storey cast iron verandah was added at the front, the windows were resashed and the house got its present front door and staircase. A single-storey verandah was built at the back and the present service wing added; the school room became the dining room and the back passage became an internal corridor. Things like ceilings and chimneypieces were replaced and the first floor was replanned to provide five bedrooms.

Although Harris died in 1891 and the house was sold three years later, it continued to be tenanted by members of the Harris family until 1957,[16] after which it was again sold, and in 1960 leased to Peanut Industries (Aust.) Pty. Ltd. and converted to a peanut butter factory. At this time most of the fittings were sold off, walls were removed, the cast iron verandahs torn off and the whole front altered to accommodate picture windows and a crude 'Festival of Britain' style porch. Steel beams were used to reinforce the structure and the back wall of the school room was taken out to allow vehicular access. Upstairs a huge gantry crane allowed the top floor to be used as a warehouse.

By 1980 when purchased for restoration by the Old Boys' Union of King's, to mark the sesquicentenary of the school in 1982, the property was subject to two tenancies — Paratune who tuned motor cars in the school room and stored spare parts in the drawing room and library; and Complete Office Supplies who rented the top floor as a stationery warehouse. The property had as well become very run down and presented a daunting picture. Everywhere was mutilation and dilapidation. Little of the detail of the colonial Georgian building was apparent and the only part of the 1890 work which would seem to have escaped alteration, was the staircase.

The significance of the building is firstly the part it played in 19th century education and particularly as the first home of Australia's oldest school. Secondly, it is likely the second oldest private house in the town of Parramatta and, for N.S.W., a rare surviving example of a two-storey verandahless Georgian house.

Where possible the restoration should reveal the earlier or pre-1890 period of the building. However, from the start it was appreciated that such were the Harris alterations that much of the 1890 work e.g. the staircase, would have to remain.

The client wanted to restore the ground floor, in particular the school room and the front rooms, and adapt the top floor as lettable office space. As well there must be custodian's accommodation to handle security and visitors to the restored part of the building. One room had to be usable as the office of the Old Boys' Union.

The first task of the conservation programme was to remove all the excrescences of the post-1960 period and then to carefully investigate the 1890 finishes and details. This work uncovered a vast amount of original detail like the gauged arches to the front windows, the elegant arched chimney wall in the library, the ceiling in the front hall, and sections of the early flagged yard and steps, plus the intact details of the 1890 front doorway and the back verandah together with its french doors. The underground water tank was also discovered[17] and details like chimneypieces and cast iron, sold off in 1960, were traced.

A major find was sections of the carved stone library chimneypiece in the ground under the 1890 verandah. Also a treasure trove of pre-1890s wallpapers was found, as well as 1830s joinery under 1890 work and so on.

With this information the conservation was refined and more of the pre-1890 period was revealed than would have at first seemed possible. Where none of the pre-1890 detail survived the late Victorian work, like the front door and staircase, was carefully preserved, or in the case of the picket fence and back verandah, reconstructed.

The corrugated iron covering the split casuarina shingle roof has been preserved together with the late Victorian arrangement of roof plumbing. So damaged by the 1960 cement render were the bricks of the facade, that the front skin had to be completely refaced by either turning the bricks or replacing them. Stone sills and the main string course are new. The windows at the front and to the school room have been resashed. Elsewhere the Victorian sashes have been preserved.

In the library the chimney wall with its cedar book shelves has been carefully restored and the stone chimneypiece repaired like a 'Greek vase', reinstated and the room painted in its original blue distemper. In the front hall the colonial cedar joinery and plasterwork has been meticu-

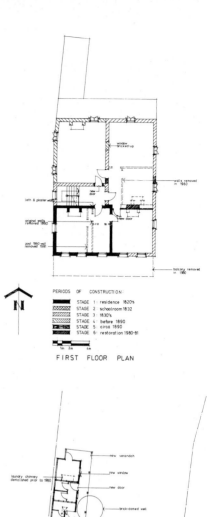

FIRST FLOOR PLAN

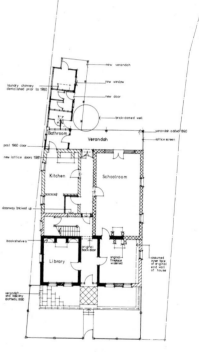

GEORGE STREET
GROUND FLOOR PLAN

Harrisford, Parramatta

Above The front wall after the 1950s cement render has been hacked off, reveals the original materials as well as the different construction to the right of the front door. (Photo by courtesy of John Fairfax and Sons)

Above right Harrisford from the rear after restoration. (Photo: Max Dupain)

lously repaired and the blocked 1860s wallpaper found there reproduced. In the drawing room an early Victorian cornice has been fitted which matches the girth of the original and an 1890 white marble chimney piece, sold in 1960, has been generously given back and is reinstated. No detail of the wallpaper survived and the top one of three found in the library niches has been reproduced.

In the school room and passage the 1890 plasterwork has been removed and the original brickwork limewashed in the colours discovered. The 'ledge door, frame, thumb latch and stock lock' mentioned in the 1832 specification has gone back at the entrance to the school room.

Behind the house, the underground tank has been dug out, its domed top reconstructed and roof water passed through it as previously. The 1860 kitchen, scullery and 1890 bathroom have been adapted as a bed sitter for the custodian. The 1890 service rooms have been adapted as lavatories and kitchenette.

Upstairs the balancing chimney stack has been put back and the surviving 1830s cedar window cases above the school room carefully repaired. In the front room, over the library, the chair rail has been reinstated and the room fitted with a donated timber chimneypiece from William Walker's now demolished house, Brisbane Grove, at O'Connell. Air conditioning plant is located in the roof space and the first floor is carpeted throughout to provide usable modern office space.

Outside the house has had its walls limewashed and its joinery painted in its late Victorian colours. A Victorian cottage garden has been devised to provide rear off-street parking for the offices and paved areas for social functions. Some of the plantings are based on historical information. The ground floor rooms have been furnished in early to mid-Victorian taste with furniture and fittings donated and purchased.

This restoration has preserved one of Australia's oldest buildings, a building which three years ago many would have argued was beyond redemption. (*Summer 1983*)

The architects, Clive Lucas Pty. Ltd., were engaged in February 1980, work commenced on the building in July and continued until September, 1982. The major contractor was Stonehill Restorations Pty. Ltd. The project won the Lachlan Macquarie award for building restoration in 1983.

Surf and steel

Rosemary Auchmuty and Peter Spearritt

Rosemary Auchmuty, whose Australia's Daughters *was published in 1978, now teaches courses on women's history and the history of London to Open University and adult education students in the U.K. From 1975 to 1977 she taught at the Wollongong Institute of Education.*

Peter Spearritt teaches urban politics and urban history at Macquarie University. His interest in the history and fate of our built environment can be seen in The Sydney Harbour Bridge: A Life *(1982) and in* The Open Air Museum: The Cultural Landscape of NSW *(1980) which he co-authored with Dennis Jeans.*

Wollongong and Port Kembla rarely hit the national news and when they do it is invariably connected with the steelworks. From the 1930s to the 1970s most of the news concerned the phenomenal growth of the steelworks, the settlements to house the new workers and their families, and other urban and commercial growth based on these twin developments. This news was occasionally tempered by strikes and industrial accidents, though the latter rarely appear outside the pages of the *Illawarra Mercury* and *Tribune*. Today Wollongong is suddenly front page news again, but this time it is because the growth has stopped. The city of Wollongong has suffered more than any other Australian city in the current recession; as the furnaces go out, the jobs go with them.

It may seem odd, at this crisis point in the city's history, to advocate a new look at its architecture — industrial, commercial and domestic — but the current recession does provide an opportunity and a breathing space for those who wish to preserve something of the city's history and beauty. With their spectacular natural settings, between surf and mountainous escarpment, and with that equally spectacular human creation, the steelworks, Wollongong and Port Kembla form a huge natural and built precinct of aesthetic, architectural and historic importance.

From Sublime Point above Bulli or from the lookout on Mount Keira (with Mount Kembla, distinctively familiar landmarks on the Illawarra skyline), the sightseer gazes out over a panorama of cliffs and surfing beaches stretching for miles to the north and to the south, over sea and lake, green pastoral lands and gaunt industrial complexes, as well as street upon street of modest suburban villas, shops, schools and hotels. In the midst of all this is the mighty steelworks. If the prevailing wind is

Right This elegant Presbyterian Church, dating from 1938, stands in stark contrast to the typically unimaginative Woolworths store that it overlooks.

Far right This classic 1938 shop front used to house Mansours drapery. Such shops have fallen on hard times. (Photos: John Storey)

kind, the air is crisp and the sky is clear, if not, then the pollution from the steelworks and associated industries casts a reddish pall over the entire landscape. Until recently there were few high-rise buildings here: the impact of postwar technology and real estate acumen upon domestic and commercial architecture is only just becoming apparent with the proliferation of home unit blocks along the waterfront and a handful of office blocks in the city centre.

Today Wollongong is the third largest city in New South Wales, after Sydney and Newcastle, and the seventh in Australia. In some respects it is not primarily an interwar city, for its greatest expansion occurred with the influx of migrants after the second world war. But it was in the 'twenties and 'thirties that it developed into a major industrial city, leading to the creation in 1947 of the City of Wollongong, incorporating the four shires of Bulli, Wollongong, North and Central Illawarra, and uniting several formerly independent villages like Stanwell Park, Coalcliff and Port Kembla.

Wedged into a narrow strip of land between the mountains and the sea — at some places the escarpment falls sheer to the sea — Wollongong has never been easy of access by land, and even today is not infrequently cut off from the outside world by flood or landslide. Until 1815 the region

could only be approached by sea, and twenty years later the decision to build a town on the site was taken primarily on the basis of its potential as a port. The name 'Wollongong' is thought to derive from Aboriginal efforts to echo the sound of the waves breaking on the shore. Primary industry prospered in the nineteenth century and coal-mining commenced at Mount Keira in 1849 and at Bellambi in 1857. By 1859 Wollongong was considered large enough to be incorporated as the first country municipality in N.S.W.

Secondary industry was encouraged by the State government which had undertaken, in 1898, to construct a deep sea harbour at Port Kembla, 12 km south of Wollongong, and to sell off adjacent land to private industrial interests. This site was chosen because it was near the Illawarra coalfields and projected rail connections from Sydney and the west. Up till 1916 the only rail access to Port Kembla was through Unanderra along the resumed colliery lines. In 1916 a direct link was opened between Wollongong and Port Kembla, and a loop line through Port Kembla. The rail line from Port Kembla to Moss Vale, no longer in operation, was not completed till 1932.

Locally quarried stone was used to build the breakwaters which enclosed the port. By 1938, four loading jetties were in existence, and an inflammable liquids berth was added soon after. The Department of Public Works had already resumed 496 acres of land at Port Kembla in 1900 for the projected harbour improvements. This proved insufficient for the industrial expansion of the area, so in 1913 a further 1,470 were purchased and, by 1938, 1,966 more; these were sold off piecemeal to the various industries which moved in. The first secondary industry established at Port Kembla was the Electrolytic Refining and Smelting Company, the largest industry of its kind in the Empire, which purchased 52 acres from the Department of Public Works in 1908 to set up a factory to refine copper, gold, silver and platinum. It was followed by Metal Manufactures (copper cables and tubes) in 1917, Australian Fertilisers in 1920, the Ulladulla Silica and Firebrick Company (later Newbolds) and the South Coast Timber and Trading Company. The Electrolytic Refining and Smelting Company generated its own electric power which it extended to

Left This advertisement, which appeared in the *Australian Women's World* in October 1929, urged travellers to come and play in the sunshine, and stay at the Hotel Ranelagh.

Right The main street of Port Kembla, looking north towards the brooding presence of the steelworks. Most of the buildings are recognisably interwar, with the notable exception of the 1880s hotel halfway down the right-hand side of the street. (Photo: John Storey)

the residences of staff and employees and later to the streets of Port Kembla. However in 1920 the government set up an additional power station, and electricity became available to Wollongong and Bulli by the mid-'twenties.

The steel industry came to Port Kembla in 1926. Steel manufacture had begun in Australia in 1900 with the installation of a small furnace at Lithgow. This was taken over in 1908 by Messrs G. and C. Hoskins who, after considerable expansion, sought a new and more accessible site for their operations and settled on Port Kembla. A new company, Australian Iron and Steel, was floated, and construction of the first blast furnace and wharf commenced in 1927 on a site near Cringila Station. Further expansion was held up by the depression but the period after the company's takeover by BHP in 1935 saw steady growth which accelerated after the war. By 1939 Port Kembla had almost equalled Newcastle's annual production of pig iron and was catching up in other areas. Within twenty years it was to outstrip Newcastle to become the most important heavy industrial town in New South Wales, producing more steel than anywhere else in Australia.

In 1936 AI&S purchased an additional portion of government land around Tom Thumb Lagoon. According to the terms of the agreement the company was obliged to expend one million pounds on the erection of permanent and fixed improvements. Among the most important industrial structures to be built in the precinct were a second blast furnace, new basic open hearth furnaces, an 800,000 gallon tank for water storage, and 72 new by-product coke ovens. These were lit in a two minute ceremony on 19 October 1937 by various officials of the company including C. H. Hoskins, General-Manager of AI&S, Essington Lewis, General Manager of BHP, and H. G. Darling, Chairman of Directors of AI&S. The ovens were then left to heat up for nine weeks, the first coke being produced in January 1938.

A new chemical and metallurgical laboratory, described as one of the finest in Australia, was erected in 1938 alongside the enlarged Administrative Offices. Both buildings enjoyed an interesting view over the works below, with mountains and ocean beyond, 'the cream walls and red tiled roof contrasting pleasantly with the green hills in the distance'. So said the *BHP Review*; like most senior management, BHP executives liked to distance themselves both spatially and aesthetically from their workforce in the adjacent industrial plant.

The town of Port Kembla fell within the Central Illawarra Shire, whose Council was empowered in 1936 by a government-appointed Port Kembla and Environs Planning Committee to construct civic improvements necessitated by the growing population, which reached 5,000 in that year. (The population of Wollongong was 13,000.) These included water, sewerage and drainage provisions; gas and electricity services; roads and footpaths; hospitals, schools and parks. Future development was planned in residential, industrial and commercial divisions, and after the economic setbacks of the early 'thirties, building went on apace. Among the most notable Council building projects in these years were the Olympic standard swimming pool (1939) and the bridge over the entrance to Lake Illawarra, making a direct road link between Wollongong and Port Kembla.

Most of the industrial structures built on the Port Kembla site by 1938 are still there, though a number have of course been modified to take account of technological changes in production processes. Today, the highway that links Wollongong to Port Kembla goes straight through the steel works, affording tourists and inhabitants alike a striking view of a twentieth century artefact of remarkable proportions. While Australian industrial archaeologists now wax eloquent over Lithgow's nineteenth century relics, and movie audiences gasp at the majesty of the abandoned Glen Davis shale oil mine, functioning industrial plant has few admirers. To many workers the Port Kembla plant represents bad working conditions, pollution and lost job opportunities. To BHP, fearful of the future profitability of steel making in Australia, the plant represents technologically outdated and expensive, labour intensive means of production. Yet the Port Kembla complex is also an architectural and industrial precinct with much to tell us about the history of technology, labour relations and industrial design in this country.

How does one develop preservation criteria for such plants? Is it practicable for major items of machinery to be moved from the site for preservation, is it inevitable that technologically outdated buildings and structures be demolished? Until recently few people, let alone governments, were prepared to challenge the notion that domestic and commercial property owners should have almost total control of what they did with any structures on their property. But today some Australian States have conservation legislation which directly impinges on those rights, though the legislation is often not enforced (as in the case of the Rural Bank building in Sydney). Some of our State governments have been brave enough to prevent demolition of domestic and commercial property, but few have carried this intervention as far as industrial property, unless it is abandoned or patently uneconomic. Brickworks are more likely to be preserved than blast furnaces, though even with brickworks high land values often defeat conservation initiatives. There are no easy answers to the preservation questions posed here, but it would be good to see some debate about the fate of outdated industrial plant and of the rights of citizens to view it.

Little of its interwar character remains in the Central Business District of Wollongong, although parts of the streetscape are relatively unchanged. Most of the shops and offices built in that period and still standing (for instance, Waltons — formerly Marcus Clarks — which date from 1920; parts of Waters, 1925 and 1930; and the Wollongong Gaslight Showrooms at 126 Crown Street, 1938) have been given extensive facelifts, often execrable ones, so that the original appearance is now masked behind Besser blocks and metal screens. But the black glass and chrome of H. Parsons, Funeral Director, is genuine Art Deco.

It is a great misfortune that no theatres or cinemas survive in central Wollongong from movieland's heyday. Even the refurbished Crown

Theatre whose fabulous foyer featured 'sunken wells with concealed strip lighting, as well as modernistic pendants' has vanished without trace. Hotels have fared better. The Illawarra and the Grand on Keira Street, the Oxford at the east end of Crown Street and the Crown at the west end all exhibit Art Deco influences in the lettering, glass and chrome work, decorative brick work and interior design. The Crown was built in 1927 in red brick with cream plaster and tiles to complement the railway station opposite.

For another memorial to interwar culture, see 'Lang's Corner' on Crown Street, especially its wonderfully appointed cafe. The Bowling Clubhouse opposite Woolworth's giant supermarket in Church Street is 'twenties-colonial, dignified in red brick. Occupying a fine corner site across the road from Woolworths stands St. Andrew's Presbyterian Church. It was erected in 1938 to replace an earlier building on Crown Street which was demolished in the previous year. Considered by *Building* magazine to be 'a splendid example of modern ecclesiastical design in brick', it was built by H. W. Thompson and Co. to the plans of Adam, Wright and Appleby. The design, as *Building* magazine pointed out:

... has obviously been inspired and influenced by the Romanesque. The manner in which the tower has been treated in a series of unfilled arched openings imparts a light and graceful effect.

The Church of Christ nearby in Kembla Street is treated similarly in red brick.

Wollongong Hospital necessarily expanded during this period with additions throughout the 1920s and a comprehensive remodelling scheme in 1930. The new Nurses' Quarters, completed in 1939, were in the 'modern style', with cream facing bricks and terracotta tiles, and a superb view over the coastline. The Trades School was erected in 1933 under Unemployment Relief conditions. It was brick on reinforced concrete, with an iron roof. The first step towards higher education in Wollongong was taken when workshops for a Technical College were erected on a new site behind Wollongong High School. Today a corridor of higher education extends from the Technical College to the Institute of Education, set in the foothills of Mount Keira, with the University in between. The Institute and the University have recently been amalgamated and they certainly form a higher education precinct but because they were not planned together they are architectually at odds with each other.

Wollongong's main street runs from west to east into the sea. An unusual landmark at the sea end is the new lighthouse erected in concrete on the headland overlooking Wollongong Harbour in 1937 to replace the old lighthouse dating from 1871, which still stands but is no longer operative. When seabathing and sightseeing became simultaneously fashionable pastimes early in the twentieth century, Wollongong — offering both — became a popular tourist resort. Alluring brochures and cheap excursion fares drew people to the South Coast and in the interwar years most of Wollongong's sixteen fine surf beaches were provided with ocean baths, dressing sheds, surf clubhouses and refreshment kiosks. The

northern beaches, being most accessible to Sydney-siders, were the most popular and the two-hour rail journey cost 4/6 return. Special trains ran at the weekends, and most were packed. Localities which had been little more than a railway station, general store and a handful of houses at the turn of the century acquired the status of fashionable and important resorts in the interwar years and a number of guest houses were built.

Far away inland on the highway between Bowral and Wollongong the remarkable Hotel Ranelagh was opened in 1928. A gigantic country hotel perched at the top of the Macquarie Pass, with sweeping views over the Illawarra valley to the coast, it was unusual for its time being 'planned from the outset for the comfort and convenience of the tourist', not for the hard-drinking public. The architectural style, observed the *Construction and Local Government Journal*, was 'dignified and simple ... nevertheless happy in appearance with spacious balconies on the gabled front'. Even when first opened it was old-fashioned in appearance and in function, but although rarely used for its original purpose (it served for a time as a convent) it has survived to this day, through many changes of tenancy, to become a conference centre. It is still possible to break one's journey for a traditional lunch or tea at Ranelagh, served in the gracious setting of a bygone era.

The 'twenties and 'thirties are best evoked by the main streets of the original industrial suburbs like Port Kembla and the still older villages to the north which, united in the shire of Bulli, grew and prospered in the post-industrial boom. Here, since rebuilding and additions have been less common and less vigorous, much of the suburban fabric of the interwar

Two lighthouses, marking different eras in Wollongong's history, overlook the city's boat harbour. In the foreground are the Continental Baths, opened in 1922. (Photo: John Storey)

period remains. If you look down the main street of Port Kembla, with the blast furnaces roaring and smoking behind, the 'twenties and 'thirties come alive. Most of the main buildings were built between the wars: the shops, the offices, the factories, the railway stations, even the police station. The dates are still emblazoned on their parapets: 1922, 1924, 1937. The main pub — the Steelworks at the corner of Wentworth and Jubilee — is recognisably interwar. This streetscape is of such architectural and historic importance that action should be taken immediately to preserve the most notable buildings and the 'feel' of the street. There is a good case for placing height limits on any new construction. The first school buildings date from 1916. There is plenty of interwar brick and weatherboard housing to be seen in the streets alongside the steelworks. The suburb grew so fast in those days that the Public Works Department was forced in 1939 to throw up an estate of temporary housing for workers and the unemployed of the district. There were huts for married folk and dormitories for single men, with communal facilities, expeditiously constructed in weatherboard with corrugated iron roofs. Some of these are still there. The pattern of cheap interwar housing, both brick and weatherboard, spreads into surrounding suburbs like Coniston and north to Crown Street, often interspersed with postwar Housing Commission settlements.

It was in the northern suburb of Thirroul that D. H. Lawrence stayed in 1922 while he was working on *Kangaroo*, in the nondescript bungalow called 'Wywurk' perched on the escarpment above the sea. The house is still there, exactly as described in the book. Thirroul and its neighbouring suburbs of Woonona, Bulli and Austinmer are really interwar districts, though their origins stretch further back; their Post Offices and Public Schools date from this time, as do Bulli Council Chambers (Architects: Kaberry and Chard, 1928), the Police, Fire and Ambulance Stations, Thirroul Telephone Exchange and Railway Institute, and several churches and banks. The Royal Theatre at Bulli and the Kings and Palais (now Hardie Rubber) at Thirroul were creations of the 'twenties. Woonona's interwar cinemas include the Vista and the very stylish Spanish style premises occupied by John Thorne Imports.

Additions were made to Bulli Hospital in the early 'twenties and again in 1933–35, with some further building 'in modern design' in 1938. In 1935 the Woonona School of Arts, built at a cost of £2,000, was opened by the General Manager of the Bellambi Coal Company. At Austinmer, the 'Brighton of the South' (even, later, the 'Riviera of New South Wales'), the surf club rooms and dressing sheds were opened in 1930 by the new Governor, Sir Phillip Game, and electric lighting was installed on the beach for night bathing. These inauspicious but pleasantly placed buildings should be preserved.

The best examples of interwar housing are often to be found in these northern suburbs, but you have to look for them among the debris of modern development. The search is not so difficult closer into town: in the modest bungalows of North Wollongong, in the red brick flat blocks and white or yellow international style residences at the beach end of

Wollongong, in the architect-designed homes in the exclusive heights overlooking the city. The trademarks of the era are there; you can pick them easily — figured brickwork, ziggurat motifs, horizontal-with-one-vertical-contrast layout, glass bricks, portholes, chrome — but so subdued in their use, so understated, that nothing stands out particularly. Wollongong was never upmarket; its inhabitants are followers, not leaders, of fashion, and prefer to be safe rather than adventurous. There is the odd exception, of course. High-trees, at 33 Mangerton Road, is listed by the Royal Australian Institute of Architects because of its fine Art Deco interior. It was built by Hugh Brit in 1935.

Wollongong is exceptionally well supplied with war monuments, each suburb having sent its complement of sons to serve in the Great War. One of the earliest built, the Soldier's Memorial at Thirroul, was unveiled on Anzac Day in 1920. It was followed by Austinmer's War Memorial Fountain (1922), a conventional plinth at Corrimal (1922), and Bellambi's simple monument, on the station platform and visible from passing trains, which was opened by the Premier of New South Wales, Sir George Fuller, in 1923. The florid War Memorial in front of Wollongong's modern Town Hall in Crown Street also dates from 1923 and was opened by the Governor General. It comprises a rough-hewn arch adorned and surmounted by various symbols of peace with the names of the fallen, along with some lines by Henry Lawson, inscribed on inlaid tablets. In 1924 Woonona-Bulli paid £400 for a 'freestone column on a trachyte base', erected on the corner of Hopetoun Street and the Princes Highway. A sad portent of the future was the Soldier's Memorial Hall in central Wollongong, which was not completed until 1938. Described in *Building* magazine as 'appropriate and attractive, in marked contrast to the ordinary buildings serving similar purposes that are too often seen in other parts of the country', it featured the Rising Sun emblem of the A.I.F. as a spandrel to the central window and a bas-relief depicting battle scenes on the 'balconette' over the main entrance.

The Illawarra coastline has long been appreciated by those who have had the good fortune to live beside it. Where once upon a time a railway line ran along North Wollongong Beach to the boat harbour, now pedestrians stroll past pavilions, dressing sheds and Continental Baths built in the interwar years. Just past the baths are the North Wollongong dressing pavilion and the Surf Life Saving Club's premises. Together these buildings, with their spectacular ocean setting, make an architectural and aesthetic precinct of considerable grace. Permanent conservation orders should be placed on all the structures in this pedestrian precinct. On the motor road above, the few fortunate inhabitants of interwar homes have been joined, in the 1960s and 1970s, by a new class of medium-rise luxury flat dwellers. From the higher units in these blocks the view in every direction must be hard to rival: north along a coastline of golden sand merging into jagged escarpment; east, over the glittering Pacific; west, to green pastures and blue hills; and south, towards the terrible beauty of the steelworks, black and red and grey, the lifeblood and the reason for Wollongong. (*Winter 1983*)

Australia in Trust

Left 'The marine villa', Lyndhurst, in its original idyllic setting on the shores of Blackwattle Bay in 1837. (Detail from a pencil sketch by Emily Anne Manning in the Mitchell Library)

Below Lyndhurst at its worst.

Lyndhurst — a battle won

Clive Lucas

Clive Lucas, O.B.E., is a Sydney architect well known for his restoration of historic buildings, including Elizabeth Bay House by the architect John Verge. Here he writes about the saving of another important Verge house, Lyndhurst, something which seemed virtually impossible 10 years ago.

'Mr. Bowman has a very splendid residence erecting at Glebe farm ... and no expense has been spared in rendering it a special model', reported the *Sydney Gazette* on July 22 1834. In 1972 the Sydney press reported the fate of Dr. Bowman's splendid residence with such headlines as 'Dog racing put before old mansion', 'History stands in path of expressway', 'Fight to save another battered piece of history' and with Members of Parliament describing the house as 'of little interest' and 'decrepit'. The north-western distributor was planned to pass through the site of Lyndhurst and the house, battered by the elements and mutilated by man, had been forgotten as the historic house it was.

Lyndhurst was built for Dr. James Bowman (1784–1846), the principal Colonial Surgeon, on 36 acres he acquired in 1833. John Verge (1782–1836) was the architect. In 1823 Bowman had married Mary Isabella, the second daughter of Captain John Macarthur, who brought with her a considerable dowry. Like the Macarthur house at Camden, Lyndhurst is named for one of the Macarthurs' influential English patrons, Sir John Singleton Copley (son of the Royal Academician of the same name), first and only baron Lyndhurst (1772–1863).

In April 1833 Verge selected the site overlooking Blackwattle Bay (one early description of Lyndhurst termed it a 'marine villa') and designs were prepared during May and June. Construction commenced at the same time with the sinking of wells and the excavation of the cellars, and the house reached completion in November 1836.

The furnishings for the main apartments at Lyndhurst were supplied by the West End cabinet maker Henry L. Cooper of 57 Conduit Street and sent from London in 1835. There were fitted Brussels carpets and Brazilian rosewood and mahogany furniture. In the dining room there were twenty Spanish mahogany chairs and in the drawing room the upholstered rosewood furniture was covered in drab merino damask trimmed with blue silk gimp with curtains to match surmounted by 'richly carved cornices finished in matt and burnished gold'.[1]

No expense seems to have been spared; the house was finished in every detail and well furnished. This contrasts with Camden Park where as late as 1841 Mrs. Bowman's sister-in-law, Mrs. James Macarthur, stated that although the rooms were well stocked with furniture 'for the most part ... all of English manufacture ... the walls were uncoloured and there were no carpets or curtains.'[2]

The design of the house in many ways resembles Verge's earlier Camden Park, which was built for the Macarthur family between 1831 and 1835. While, however, it is similar in detail, the arrangement of rooms is quite different and the house is larger in scale.

Lyndhurst has three main elevations with a long service yard at the back. The design was undoubtedly influenced by Camden Park and, as James Broadbent has suggested, that house was most probably a synthesis of ideas culled from a wide variety of pattern books collected by John Macarthur. The Italianate character of the design undoubtedly derives from Robert Wetlen's *Designs for Villas in the Italian Style of Architecture* of 1830 which is still in the library at Camden Park.[3]

Not only had James Bowman spared no expense on the house but neither had he on the grounds and its appurtenances. As well as the large service yard and balancing wings at the back of the house, there were stables designed by Verge and elaborately laid out pleasure grounds, shrubbery and kitchen garden. The estate was well known, described first by Thomas Shepherd in his *Landscape Gardening in Australia* published in 1836, and later in J. C. Loudon's *Encyclopaedia of Gardening* as an example of antipodean gardening.

Lyndhurst's use as a family home was shortlived. In 1842 Bowman sold the property to his brothers-in-law James and William Macarthur with whom he had become deeply and foolishly involved over the Australian Agricultural Company. By 1847 James and William Macarthur had become heavily indebted to the Bank of Australia which accepted Lyndhurst, with other of their properties, as part settlement of their debt.

In 1852 the Bank sold the property to the Roman Catholic Church for St. Mary's College. The historical significance of Lyndhurst is as much due to its association with Church education as with the Macarthur Bowman family. From 1847 to 1849 it housed St. James's College, Australia's first theological college. During this period it produced several distinguished native-born clerics and its classical studies gave Australia's first Anglican bishop, Bishop Broughton, hope that it would play a part in the then merging university movement. Its churchmanship was, however, considered too high and this involved it in a major crisis which led to its demise.

The Roman Catholic St. Mary's College shared a similar fate. It taught secular pupils while the English Benedictine community which provided the teaching staff formed a regular order. Until the late 1860s, its elaborate classical curriculum and high scholarly standards made it the most successful school in the colony. Thereafter it declined. The competition of country Catholic schools, the Irish dislike of English Benedictinism and a noticeable slackening of morale in the Benedictine community weakened its popularity. St. Mary's had been formed by Archbishop Polding and following his death in 1877, his successor, Vaughan, closed the College, subdivided the land into small blocks and auctioned the estate at sales held in 1878 and 1885.

Between 1852 and 1878 at various stages, additions were made by St. Mary's College until there was a long range running south from behind the service wings. Except for the end of this wing, which still exists, all the school additions together with the original service wings and stables were demolished prior to the 1878 sale.

At the auction held in September that year the house was sold to Morris Asher (1818–1909), a businessman and Parliamentarian. Asher occupied the house at first but in 1882 leased it as a lying-in hospital. In 1890 he converted the building into three dwellings at which stage the verandahs and portico were demolished. One of the dwellings (no. 59 Darghan Street) became the 'Lyndhurst Private School' which continued to function until 1908. The property changed hands several times after Asher's death in 1909, first to Joseph McFarland and then in 1925 to Aubrey Bartlett, whose widow sold it to the Department of Main Roads in 1972.

Lyndhurst sank further into oblivion, hidden by the terraces of houses which went up following the sales of 1878 and 1885, and itself disguised as three terrace houses in 1890. Its successive uses degraded and mutilated it further. It was used for, among other things, a laundry, a cabinet

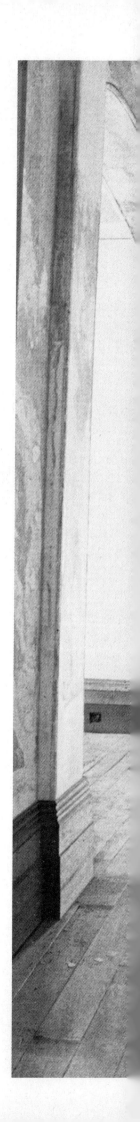

Left The cross hall with its original plaster vaulting.

Top The drawing room in 1982.

Middle Lyndhurst from the south east in January 1982.

Above left The front hall in 1982.

Above right A typical doorcase. The detail is quite close to that at Camden Park. (Photos: C. Lucas)

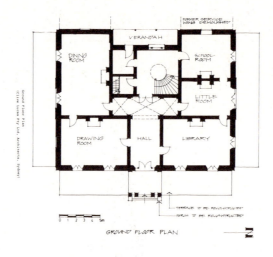

The dining room french doors before restoration. (Photo: I. Stapleton)

maker's workshop, an icecream factory, broom factory and printing works. Skillion roofed sheds leant against the once handsome walls and it became more unkempt and neglected. Until 1973 it had been occupied but the Department of Main Roads evicted the tenants preliminary to its demolition and things got steadily worse.

It had always seemed to me that despite the terrible vicissitudes Lyndhurst had suffered, a great deal of the original house remained and that its restoration was quite feasible. In 1972 I consented to chair the *Save Lyndhurst Committee* to try to get the route of the expressway changed and Lyndhurst protected for eventual restoration. A large campaign was mounted, supported by the National Trust, but despite every effort there was little real hope that the route of the road would be changed and arguments were put up which made its restoration seem impractical anyway. The pressure of the campaign did, however, prevent the house being demolished, although how it survived is a miracle.

The house was unoccupied, it became further vandalised, the roof finally rusted through, fires were lit by various derelicts who squatted in the building, the southern ridge was burnt out, water poured in through the rusted box gutters and broken windows and doors. It became the home of pigeons and vermin of every description.

The Labor government which came to power in 1976 put a halt to inner city expressways and indeed one of the first questions the new Premier, Neville Wran, put to the National Trust deputation which met him in June, was what did they want done with Lyndhurst? A miracle had happened.

Despite the Premier's undertaking that the house would be preserved, it took a further three years before the commission came from the Heritage Council of New South Wales and work could commence on Dr. Bowman's splendid house. The work has been staged in such a way as first to hold the house and then to reveal its major architectural character. At the end of last year some $175,000 had been spent. This has given the house a new corrugated iron roof, made it structurally sound, provided a caretaker's flat, restored all the external joinery and restored the handsome entrance hall, drawing and dining rooms to their original form. Lyndhurst is now safe and recognisable again as a fine house with a stunning interior. (*Winter 1982*)

Edmund Blacket's church architecture

Joan Kerr

Dr. Joan Kerr is lecturer at the Power Institute of Fine Arts, University of Sydney. She was guest curator for an exhibition in January 1983 celebrating the centenary of Blacket's death, sponsored by the National Trust of Australia (N.S.W.), the Mitchell Library and the University of Sydney.

When establishing himself in New South Wales in the early 1840s Edmund Thomas Blacket had the help of two great admirers within the Anglican church: the Bishop of Sydney, William Grant Broughton, and the Highest Churchman in the Colony, William Horatio Walsh, rector of Christ Church St. Lawrence. Blacket got to know Walsh through completing his church for him, but he brought a letter of introduction to Bishop Broughton from England when he arrived in Sydney in November 1842. It was from a surprisingly eminent source, for although the letter itself has been lost we know what it said. Broughton summarised it in a note he wrote to Governor Gipps:

Mr. Blackett (sic) is a gentleman who comes out with the sanction of His Grace the Archbishop of Canterbury, and strongly recommended as conversant with all that is essential to the successful management of schools for General Education.[1]

With such a reference it is no wonder that Broughton appointed Blacket 'Inspector of the Schools in connection with the Church of England' in the Colony on 1 January 1843, less than two months after his arrival. This note also suggests that Blacket had done more than sketch medieval buildings and get married in 1842; he had obviously fitted in some experience in school management and possibly design. Perhaps some of the six hundred pounds he arrived with was earned in this way. It almost seems as if Blacket was preparing for a colonial career.

Yet we know that this was not the case. In 1849, when Blacket was the official Colonial Architect of New South Wales, he wrote to his brother Frank in London:

I have a strange variety of matters, and I find that every single scrap of knowledge that I ever picked up anywhere is of service to me. I only wish I could have forseen my present career during the year I wasted in Misery and London before I sailed, but perhaps even that was of use, as it certainly made me very contented with all my lot ever since.[2]

Sadly, he didn't expand on why this year was such a misery. It is generally thought that it was because his father disapproved of Sarah Mease, the Yorkshire girl he finally married at the end of the year and immediately brought to the other side of the world.

In Sydney, Bishop Broughton had been waiting for someone like Blacket for years: a gentleman, a churchman, a medieval architectural enthusiast, a draftsman who could draw details that an illiterate builder in the country could get right without supervision (or almost right), and a reliable estimator of costs and time of building programmes. Until Blacket arrived Broughton frequently had had to design his own churches and get James Hume to draw them up for him. Hume was, said Broughton, 'a Presbyterian quite ignorant of church architecture.'[3] Once Blacket appeared, Broughton gave him almost every bit of ecclesiastical work in the Colony and the bishop was never disappointed in the results. They were all, as he said of Wollongong, 'well designed and well built'.⁴

Yet Blacket had put up his 'elegant houseplate' inscribed 'Architect and Surveyor'⁵ with some bravado and a little private trepidation, as he confessed to his brother Frank in 1843, for he was not really either. Still, as he said:

*There is nothing to be gained here by hiding one's talents in a Bushel and if you were to see the impudence of some of our professionals you would think me very reasonable in pretending to something I had studied instead of setting up for Doctors, Musicians, painters etc. by folk who are preposterously ignorant of their new invocation.*⁶

He had also written:

*I have however now fallen into such circumstances where I can indulge my Architectural propensities without one twinge of conscience, and hope to have a great hand in improving the taste of the discerning Public upon Ecclesiastical Architecture. I have already succeeded marvellously and although I do not officially bear the title, yet I hear I am pretty well known as the Church Architect. Certainly all the work of the kind begun here since I came has fallen to my share.*⁷

This manna had largely fallen from the hands of Bishop Broughton.

Blacket's chief architectural disciple and lifelong friend in the Church, however, was Canon Walsh at Christ Church. Walsh's church had been designed in 1840 by the builder-architect, Henry Robertson, and was well under way by 1843, but Walsh became unhappy about the result after being exposed to Blacket's new ideas. In a lecture Walsh gave on the church architecture of New South Wales in England in 1851, he explained how Christ Church had begun as an architectural disaster — speaking as a converted Victorian medievalist.⁸ The buttresses were starved, the mouldings were too shallow, the west door was huge, and the lancet windows were too wide. There were blunders in construction and the proposed internal arrangement was impossible. The best thing that could be said for Robertson's building, Walsh thought, was that 'the masonry, as far as it went, was sound'. Then Blacket was appointed and did 'the best he could out of such unpromising materials'. He finally gave the interior a truly 'church-like character'. He designed 'correct' medieval-styled open seats for the congregation, instead of box pews, and stalls for the choir set in two sideways blocks between nave and chancel. He added a 'handsomely carved' lectern, pulpit and font, put in a false cedar roof suggestive of a medieval one, shortened the east windows, and added a northern vestry so that the original eastern vestry could be opened out to form a proper chancel.

It is clear that these liturgical changes were not a minister's idea, although not normally coming within the province of an architect. In 1840 Walsh had been quite happy with Robertson's plan,⁹ so it was Blacket who changed this Gothick box into a correct Early Victorian church. His open seats were the first medieval ones seen in Australia, with large 'poppy-heads' on the ends of the benches.

All such medieval-inspired details that Blacket introduced into Australia had to come out of books, or from sketches that he had made in England. Australia was singularly lacking in genuine medieval examples and the whole point of the exercise was to get all these details 'correct' — including exact mouldings and poppy-heads. To invent details was to go back to Robertson's 'starved buttresses' and to 'Carpenter's Gothic' rough approximations. At the least, invention or originality in design meant you were putting your own creative importance above hundreds of years of Christian worship and the values of a truly Christian society. Blacket was a lifelong Romantic. His ambition was to recreate this perfect Christian past on Australian soil.

At Christ Church Blacket also introduced the medieval practice of

Elevation plan for St Stephen's, Newtown, 1871. Ink and watercolour, 70 x 49 cm, Mitchell Library D199–1,67. This was Blacket's most successful tower, set in an unusual position between north aisle and transept.

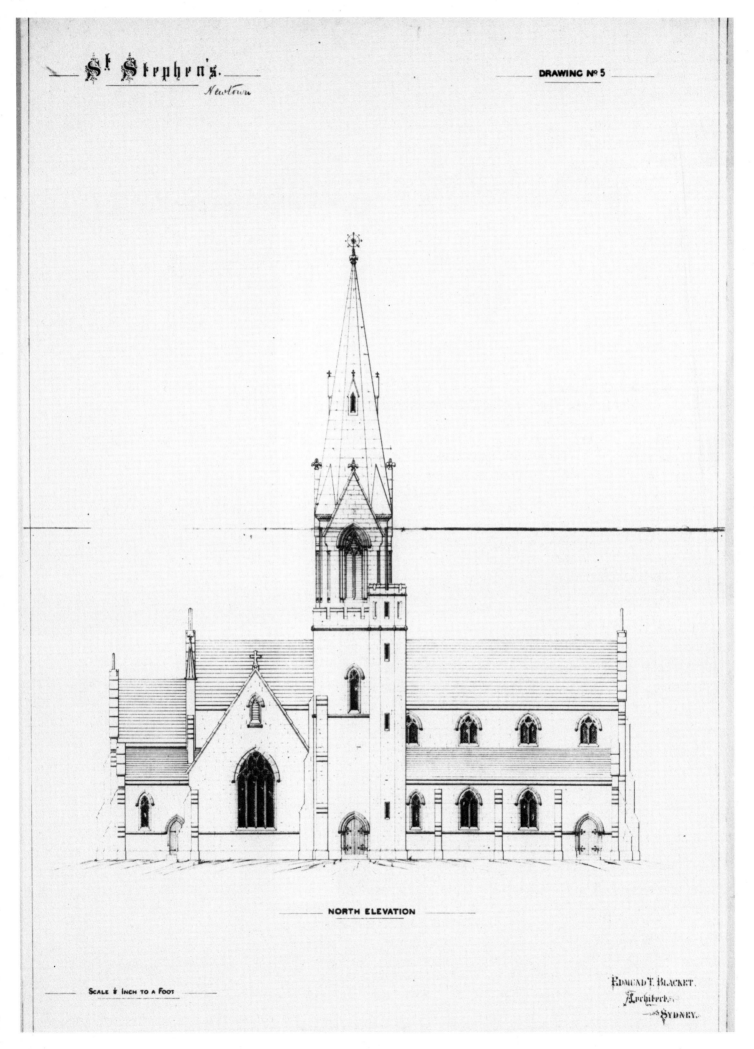

elevating the chancel several steps above the nave and making its focal point the 'altar' — a word that offended the Low Churchmen of Sydney who called it a 'communion table'. (Previously the focal point of an Anglican church had been the triple-decker pulpit. High box pews with doors were arranged around it.) Blacket also added an external cross on the roof of Christ Church, because English medieval parish churches had them. Walsh preached in his surplice. All these things upset a lot of Sydney Anglicans, one of whom wrote to the Sydney Morning Herald complaining of these 'Puseyite' innovations at Christ Church:

Of little use will be the endeavour they are now making in England to stifle the Archfiend, if he be allowed to sow his Jesuitical doctrines even in this distant portion of the Empire.[10]

Stone crosses were thought especially offensive symbols of the Archfiend i.e. 'Papists'. Clearly, architectural innovation in churches could be dangerously controversial. Nevertheless, Blacket persisted in introducing what he called '*decency* in churches as well as economy'.[11]

The most architecturally significant of the innovations at Christ Church was the matter of the choir being 'disposed for antiphonal singing': that is, the seats for the choir were placed in the nave between the people and the sanctuary and arranged facing one another. The men sat on one side and the boys on the other. (Women didn't sing in Medieval Revival churches, although they had done so earlier when the choir was up in the back gallery with instruments like fiddles rather than an organ. In Medieval Revival churches, like Blacket's, the back gallery became the place for the Sunday School children.)

This choir arrangement would appear to be right outside the province of an architect, but I think it had to be Blacket's innovation in the Antipodes, although justified by both modern and medieval English authority. The first English Anglican parish church to revive this choir position in the nineteenth century had been St. Peter's at Leeds in 1841. In 1841, Blacket was, for a time, working as a surveyor on the Stockton to Darlington railway and was in the area. His wife's family came from Yorkshire, not far from Leeds.

Blacket was also an enthusiastic church organist and very interested in musical improvements. He played regularly in the pro-cathedral in Sydney and in his youth had built an organ for the Wesleyan Church at Stokesley in Yorkshire. In Australia, he had control of the installation of English organs in dozens of his buildings, including St. Stephen's at Newtown, the University of Sydney and St. Andrew's Cathedral. Leeds parish church had received great publicity when it was first opened and the advantage of this choir position for music and singing stressed. Blacket must have visited Leeds and been inspired by its arrangements.

By 1843, The Cambridge Camden Society, an extremely influential body of High Anglican lay churchmen obsessed with reviving medieval church architecture, had decided that all church choirs should be near the chancel rather than over the entrance door and 'on the floor' rather than in a gallery. They gave the Leeds innovation their *imprimatur*, but recommended that the chancel would be an even more medieval, and hence 'correct', place for the singers. Blacket continued to use both positions throughout his life.

Blacket's next church was St. Stephen's at Camperdown (1844–45), the demolished brick predecessor of the extant St. Stephen's at Newtown. It was a cheap unpretentious brick building but it had, said Walsh — who had helped Blacket get the commission — a correct long chancel to contain the singers, 'in spite of all the efforts of its church building committee to thwart the architect',[12] So, although Horbury Hunt thought Blacket was 'a timid man' when he was working under him in the 1860s,[13] it is clear that in his youth Blacket was sometimes prepared to espouse Hunt's lifelong philosophy of giving people what was good for

St Stephen's Church, Newtown. (Photo: D. Baglin)

them rather than what they liked.

Of course, Blacket needed the support of his Bishop and his priest to introduce such controversial changes. Both backed him to the hilt. Broughton wrote about Christ Church in his 1845 *Visitation Journal*, which was published in 1847 and distributed to his flock:

I have heard objections stated to some of the arrangements in the celebration of divine service, as savouring of novelty and innovation; but I am bound to say that there is no contrariety in any part of the practice to the most approved usages of the Church of England, with which I have been familiar from my earliest years; and everything is marked by such a degree of order and solemnity, that I could wish the observances of this church to be taken, if it were possible, as a model for the imitation of every church in my diocese.[14]

Another major innovation at Christ Church was a modern figurative stained glass window. Some locally-made windows had existed in churches in Australia before this, but they were necessarily hand-painted rather than fired and looked like imitation oil paintings, or were just coloured panes of window glass. None of them would have borne much resemblance to the great medieval glass of places like Chartres or Canterbury, but that was what Blacket wanted in his churches. At first he had painted imitations of stained glass for St. Andrew's pro-cathedral, which was given 700 squares of coloured glass in the east window in 1843.[15] He was so good at this that he thought of making a living from it. As he said:

... the folk who have come to see ... will hardly believe that it is not stained glass. I am certainly pleased at it, which is unusual with my own designs, but the Bishop is delighted also, which is of more consequence, and I hope that I shall be able to make a trade of it.[16]

However at Christ Church it was decided to have real English glass rather than colonial imitations. The east window of Christ Church was commissioned from William Wailes of Newcastle-upon-Tyne in 1844.[17] Blacket was clearly the major figure behind the commission.

Unfortunately, the plan for impressing the Colony with a great modern English window of a medieval type went wrong. Wailes' effort when it arrived in 1845 'savoured very strongly of something more than Puseyism'.[18] Amongst the figures depicted was the Virgin Mary! Both Broughton and Walsh (and, presumably, Blacket who went with them to inspect it on arrival) thought that this was far too Papist altogether. The window was sent back to Wailes for emendation and it was not installed until 1853, although Broughton apparently saw it and approved its final state in 1852 before he left the Colony.[19] It then contained an inoffensive St. John, Christ as the Good Shepherd, and St. Peter. The 1905 fire at Christ Church destroyed most of it; a single figure remains in a small window at the base of the tower.

This debacle did not deter Blacket, who proceeded to put stained glass windows — by Hardman Brothers of Birmingham[20] — into St. Mark's at Darling Point in 1852 and 1853. They seem to have been the first imported medieval-style windows in New South Wales, although Tasmania had installed a couple before this.

Stained glass received its ultimate accolade in 1856, when Blacket began to install a complete set of windows by Clayton and Bell in the Great Hall of the University of Sydney. They had been acquired largely through the efforts of Blacket's '*firm* friend'[21] Charles Nicholson, the Provost and, later, Chancellor. Queen Victoria inspected this glass before it left England and signified her approval. There was nothing Papist about the kings and queens of England! Clayton and Bell were subsequently invited to design the glass for the Albert Memorial Chapel at Windsor.

In Australia, Blacket continued to put stained glass windows into every church he designed that could afford them; he believed in supporting local

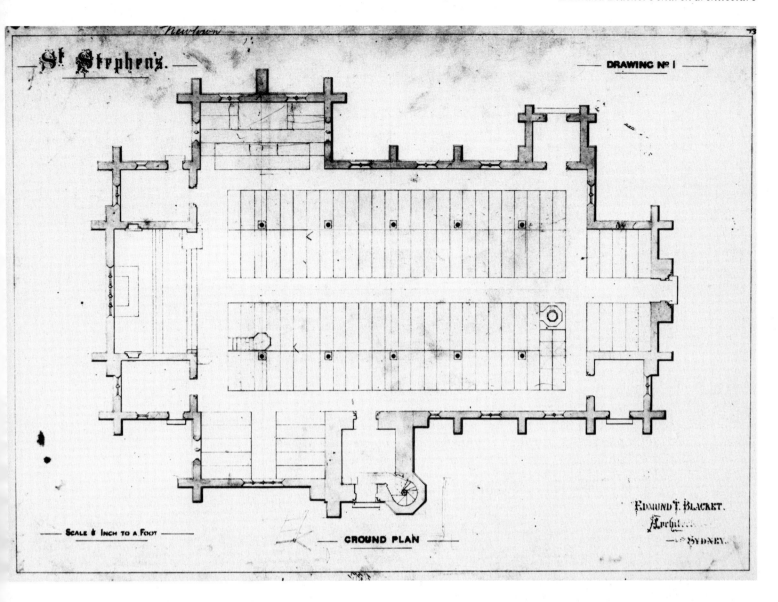

Ground plan of St Stephen's, Newtown, 1871. Pencil and watercolour, 35 × 48.5 cm, Mitchell Library D199-1,73. A typical Australian Anglican church plan of the Latin cross form.

manufacturers once they had begun production. The first professional stained-glass maker in New South Wales was John Falconer who set up kilns in Sydney in 1863 specifically to make quarries for Blacket's St. Mary's at Waverley.[22] Without Blacket's support he could never have made a living.

Cyril Blacket commented after Blacket's death that his father had always tried to encourage local craftsmen. He illustrated this by telling a story of Bishop Thomas escorting visitors through Goulburn Cathedral, a building containing glass selected by Blacket from both English and Australian makers. 'Blacket', said Thomas, 'says that windows made by Sydney artists are equal to those imported' and challenged his viewers to tell which was which. None could, of course.[23]

This is a revealing story from the point of view of Blacket's own architectural ambitions. He didn't want to be different from England; he wanted to be indistinguishable from it. As he wrote to his brother in 1843:

When you hear anyone vaunting Australia as superior to old England in any respect tell them from me that it is all my eye, that nothing *is near so good here, neither Climate, nor people, nor sail nor fruits nor anything else — not that I am complaining, far from it, I am well satisfied, having plenty of work and good health, but every person who arrives here must expect to lose a great deal in every respect.*[24]

As far as church architecture went, Blacket set out to repair this loss. The highest form of praise one could give a Blacket church was to say that it looked just like an old village church at Home. This ambition was not unique to Blacket. It was characteristic of his generation of expatriate Britons.

Blacket's ideas were rapidly accepted as perfectly appropriate for Victorian Anglican churches of any height of churchmanship. By 1848 Archdeacon William Cowper willingly accepted Blacket's medieval parish church ideal for his new church of St. Philip at Church Hill in Sydney. It had a raised chancel containing altar and choir, an external stone cross at the east end of the nave, open seats throughout the nave, and was meticulously fifteenth century in style. Yet Cowper was one of the most intransigent Low Churchmen in the Colony; he continued to wear black and armbands long after most of his colleagues had gone into surplices to preach.

By the time that the new St. Stephen's at Newtown was designed in 1871 a medieval parish church style had become the conservative choice for Anglicans, Presbyterians, Catholics, Methodists and even Congregationalists — although the Baptists were still erecting classical temples. By then, more adventurous Anglican patrons were just beginning to appreciate Horbury Hunt's 'muscular Christian' brick churches in a more original High Victorian style. Under Canon Robert Taylor St. Stephen's was the Lowest church in the Low Diocese of Sydney — a diocese that never gave Hunt any work during Barker's episcopacy and only one or two churches subsequently.[25] In some ways, it is rather ironic that Taylor should have been responsible for Blacket's most perfect realization of the fourteenth century parish church ideal.

From the late 1850s, under the evangelical Bishop Frederic Barker, the size of the chancel compared to that of the nave had somewhat shrunk from Blacket's earlier proportions and St. Stephen's has got rather a long body and a little head. However, within these modified proportions, we can still see all Blacket's necessities for a 'correct church'.

Mr. Taylor certainly would have been amazed to realise that thirty years earlier he could have been branded a Papist — or at least a Puseyite — for his stone altar and his external stone cross.[26] There is also a stone niche in the gable of each transept, another common feature in medieval churches. Like many ancient churches in England, St. Stephen's niches are empty (except for wooden louvres for ventilation). In England the statues that once occupied such frames were mostly destroyed by Tudor

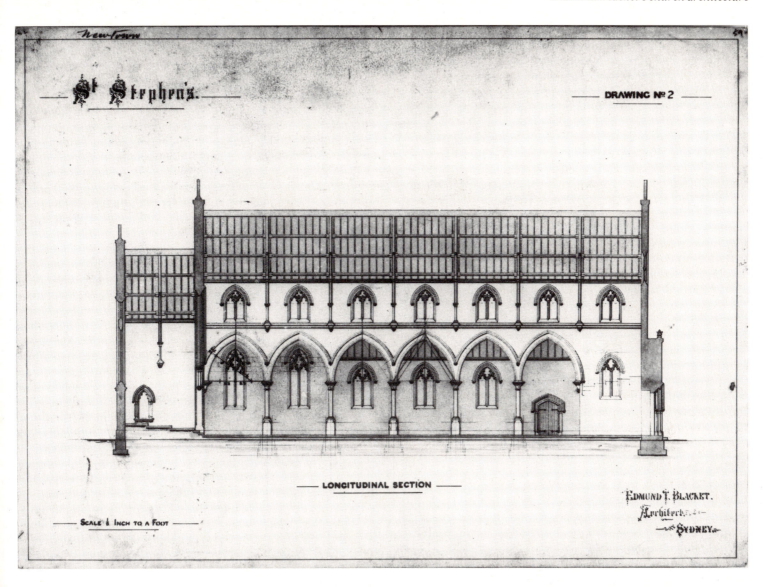

Longitudinal section of St Stephen's, Newtown, 1871. Ink and watercolour, 35.5 × 49 cm, Mitchell Library D199–1,59. A typical Blacket elevation in fourteenth century style, showing his characteristic mannerism of springing arcade arches from inside the tops of the capitals.

or Puritan iconoclasm, but Newtown's stone niches were associational, not functional, in intention. James Francis Turner, Bishop of Armidale and Grafton, wrote in 1888 that over Newtown's niche 'should be written "This shelter to be left unfurnished" for it will not see any statues'.[27] Taylor's acceptance of medieval precedent did not extend as far as that!

All features in a proper Early Victorian Anglican church were meant to have both symbolic and functional justification. That was what Blacket meant by 'decency'. For instance, the Church of God is made up of many separate souls but forms a unified whole; hence, chancel, nave, aisles, vestry and porches are separately roofed to proclaim their different functions within the unified building. The chancel containing the altar should face the east and the rising sun, symbol of the risen Christ, although even Broughton felt that this orientation had to be practical as well. Nobody wanted to lose a roof through a gale blowing straight across it. Hence normal orientation is easterly rather than directly east in Sydney.

Other liturgical modifications were also made for the Antipodes. The problem of whether the major entrance porch should be on the south side, like England, or on the north — the sunny side symbolising the soul's entrance into the Christan Church — perplexed many Australian church designers. Blacket often solved the problem by having two porches, although his major one was usually on the north. On the other hand, Early Victorian towers could, quite properly, be placed anywhere.

However, the siting of St. Stephen's tower at Newtown is unique in Blacket's work, nestling against the north transept. It is also the most powerful stone tower and spire that he ever designed. Its beauty from the churchyard seems a splendid justification for the re-creation of England in Australia, although the extreme Ocker flavour of recent vandalism in the churchyard detracts from the feeling that we have suddenly been transported to a fourteenth century village church.

All Blacket's detailing was as consistent as his planning. If he designed a fourteenth century style church then everything from the tower to the mouldings on the base of the columns was fourteenth century in style. That is another of the signs of a 'correct' Early Victorian design. The detailing is always more precise, uniform and repetitive than any genuine medieval church has retained. In England, people like Sir George Gilbert Scott made a fortune turning genuine medieval churches into what they thought *should* have been there in the fourteenth century. In Australia, Blacket just transposed our British inheritance. He made no fortune from it either, for no Australian architect has ever made a fortune from church building. When he died Blacket left £3,100, not a large amount even in 1883.[28]

Like all architects, Blacket had personal mannerisms and minor (non-medieval) lapses. At St. Stephen's the roof and the chancel arch are both rather minimal in form. They were often Blacket's weakest points in his 'Middle Pointed' churches. Still, they remain correct in period detailing. The thin necks above the capitals in the nave arcade are also typical of Blacket, but somewhat less archaeologically flawless. Bishop Turner, who had trained as an architect in his extreme youth, said that Blacket should have made his 'arch-mouldings oversail on the capitals from which they spring' for by neglecting to do so the effect was 'to give the top member

Edmund T. Blacket. Sections of seats for St Stephen's, Newtown, 1871. Ink and watercolour, 35.5 x 49.5 cm, Mitchell Library D199-1,63.

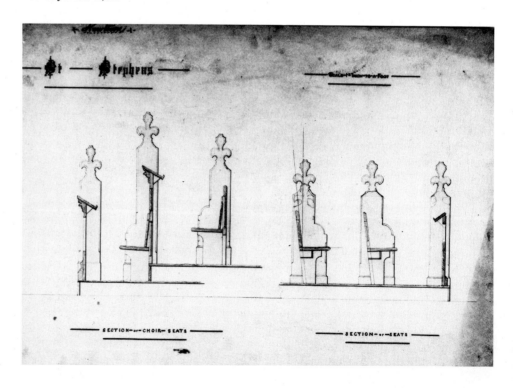

of the capital a most unpleasant prominency'.[29] Blacket's church arcades are, I think, always like this.

Turner also thought that the fillets on Blacket's window mullions (specifically at West Maitland church) were 'just twice the size that they ought to be'; while the poppy-heads on the benches in Sydney Cathedral were set too high, had incorrect detailing, and were over-large.[30] Similar criticisms could be made of St. Stephen's, although here the poppy-heads are a much more rendered-down High Victorian version of his intricate early ones.

Turner suggested that Blacket's poppy-heads ought to be cut down, but this proposed vandalism did not eventuate because Sydney diocese continued to like Blacket's old-fashioned style. Turner was writing in the late 1880s as a High Victorian criticising his Early Victorian inheritance. He would have found many of Blacket's English contemporaries 'erring' in exactly the same ways, with stone crosses on the exterior larger than medieval ones, windows of non-medieval proportions, window surrounds that were too mechanical and had keystones: 'if there is *one* thing distinctive of medieval architecture, it is that its pointed arches have *no* key stones' he asserted.[31]

Blacket got off comparatively lightly from Turner's pen. Other Sydney architects were absolutely decimated, while the poor parishioners in Turner's diocese at Tamworth had a sermon preached at the opening of their church about the architectural inadequacies of their building, to the text: 'My House shall be called a House of Prayer, but ye have made it a den of thieves'.[32] Amongst other things, he pointed out that although Christians could worship in a barn if necessary, this was no reason to build a church that looked like one.

There was some justification in Turner's criticisms of Blacket's late work, although the weaknesses were really the fault of the place, not the man. Working in isolation twelve thousand miles from one's sources resulted in a necessary dependence on published material and made it hard to develop in new directions. The same fate dogged all our nineteenth century architects and their late works are rarely as exciting as their early ones. What Blacket did was to gradually assemble a repertoire of forms and motifs from the time he arrived until the mid 1850s and these were the foundation for all his mature works. A few innovations in form and detailing were later introduced, but the basic vocabulary and grammar of his architectural language did not change.

St. Stephen's, designed when he was fifty-three, is the place where Blacket finally realised the revived fourteenth century parish church in its most perfect form. It was not trying to be an original or innovatory design, but the culmination of thirty years of experience where all the ideals of youth were perfected in maturity.

It is hard for us to grasp the concept of perfecting a common ideal when we are still living in an age when every architect wants to display his unique genius. It is also hard to understand a culture for which Britain meant everything. Yet both concepts are fundamental to an understanding of Blacket and his work. And it is sometimes difficult to accept that the arrangement and furnishings of churches were not ordained by God at the Creation. Yet the typical parish church still found throughout Australia was entirely a nineteenth century creation. It was introduced to New South Wales by Edmund Blacket, the man who has continued to be 'pretty well known as the Church Architect'. (*Summer 1982*)

Vaucluse House

Peter Watts

Peter Watts is the Director of the Historic Houses Trust of New South Wales, which is the owner of Vaucluse House and Elizabeth Bay House. He was formerly the Property Officer with the National Trust in Victoria.

Top This is thought to date from c. 1875. (Mitchell Library)

Above A view in the *Illustrated Sydney News* of 1869 uses a degree of artistic licence but nevertheless confirms some of the detail. It is the first illustration showing the fountain. (Mitchell Library)

Throughout the world the great strides which have taken place in the theory and practice of building conservation, and the interpretation of historic interiors, is resulting in a reassessment of many previously revered house museums. Even the American holy of holies, George Washington's 'Mount Vernon', is giving way to more strident colours, and painted joinery is being given the woodgrainers' touch. 'Ham House', outside London, even though one of the most intact of seventeenth century interiors, is being re-interpreted by the National Trust as research has uncovered marbled surfaces and new colours. Closer to home the National Trust in Victoria is reassessing Como following intensive research. It comes as no surprise to find its pristine whiteness, both inside and out, was unknown to its earliest owners.

In June 1980 the Historic House Trust of N.S.W. was formed and given responsibility for Vaucluse House in Sydney's eastern suburbs. The house has been a museum since 1910 when it was acquired by the State Government. Although it has come to be regarded as something of a national shrine because of its association with William Charles Wentworth, Vaucluse House was not acquired for such patriotic reasons, but rather to enable the people of this part of Sydney to have some access to the harbour frontage which was then largely in private hands. Time has enhanced the public's regard for the house and it is now undergoing a major conservation programme.

Vaucluse House was commenced in 1803 when a small cottage was built for the then owner the roguish Irish knight Sir Henry Brown Hayes. It changed hands several times until being acquired by William Charles Wentworth in 1827. Wentworth, a barrister, landowner, explorer, statesman and father, lived here with his family until his departure for England in 1853. Several waves of building activity occurred during this period, presumably to accommodate Wentworth's growing family which numbered 10 by 1853, but also to provide a suitable residence to match his growing public reputation. On the Wentworth's family departure in 1853 the contents of the house were auctioned and the house let. The Wentworths returned briefly in 1861/62 to attend to business matters and sent ahead of them a number of crates of furniture. They also extended the verandah to its present appearance and undertook other minor alterations.

Apart from the loss of a timber bathroom wing (demolished in 1916 after it was threatened by collapse through white ant activity) and a number of outbuildings, the general form of the house has not changed. There have, however, been many changes of detail. In the 1920s the then Trustees extended the castellations across the front of the house and built two turrets (in stucco) to match the two original sandstone ones. Even though this may have been Wentworth's proposal — there are no documents to prove it and the resultant effect has robbed the house of an important part of its social history i.e. the abrupt stop to building activity brought about by the depression of the early 1940s. Whether these ele-

Vauclause House

Above Paint scrapes on the ceiling and walls of the Breakfast Room revealed an earlier, but similar stencilled pattern on the ceiling, and a previously unknown stencilled frieze below the cornice.

Above right The Little Tea Room in 1982. Paint scrapes revealed the original joinery colours which were used as the basis for the redecoration of this room.

Right The Dining Room in 1982. The table is set at different times of the year and the arrangement of the rooms is changed slightly according to the seasons.

ments are to be removed is now a ticklish question. The new conservation policy for the house will certainly allow it.

Successive restoration and decoration schemes and attempts to interpret the house, had, by the 1960s, robbed the place of much of its nineteeth century detail and hindered an understanding of how the house functioned. The pantry had become a costume display room, the housekeeper's room housed a fine display of eighteenth century china, a sitting room had become a bedroom, the breakfast room was called the family dining room and so on.

Like so many historic house museums developed during the mid-twentieth century Vaucluse House had become a repository for a miscellaneous collection of objects, mostly unrelated to the house and its history. Furniture was arranged in a way which made it difficult to imagine a family ever living in the house and the general presentation tended to confuse rather than educate.

A recent period of intensive research has enabled much of the myth, which so often develops around such houses, to be dispelled, and has instead provided much information relating to the development of the house, the family who occupied and built most of it, the contents and decoration, and the development of the garden and grounds.

This has meant, for instance, that the room traditionally presented as the main bedroom will become a family sitting room — in an 1853 lease document it was described as the 'second room' and housed objects such as a secretaire, bookcases and easy chairs — hardly a bedroom it would seem. The china room will be dismantled and returned to its former use as a housekeeper's room. The fine collection of china will probably be loaned to other more appropriate institutions.

All the information gained from documentary research, together with archaeological and other physical examinations and research, has enabled a more rational plan to be developed for both house and garden and they are now being given something of the character and detail of the house in the mid-nineteenth century. Where information is missing, careful examination of contemporary photographs, paintings and books of the period are used to ensure the most objective decisions are made.

In general the interior of the house is being recreated to the period up to 1853 when Wentworth and his family sold the contents and moved to England. (Sadly we have not been able to find a detailed catalogue of that sale.) Several rooms depart from this date — always for a specific reason. The entrance hall, for instance, was sketched by Rebecca Martens in 1869 and that sketch is being used as a basis for the redecoration of the hall. Likewise the stencilled decoration of the breakfast room (found in early 1982 under layers of later paint) dates from the 1870s, thus providing a convenient room to place the 1870s dining suite which once belonged to Wentworth's son who lived in nearby Greycliffe House. Wentworth's brief return to the house in 1861/62 also provides some leeway in acquiring slightly later objects. But in general the earlier date is used, thus giving to the house a degree of authenticity which the public, no matter how little they know of the detailed history of the decorative arts, seem to respond to. Thus wallpapers, paint colours, carpets, furniture, china, paintings and so on are all of a period. The quantity of furniture and objects which is growing in some rooms may be offensive to modern taste — but better to shock and educate than soothe and misinform.

About half the new objects are being acquired in Australia by constantly scouring shops and auction rooms. The remaining objects are being bought in Britain by dealers who are intimately involved with the refurnishing programme and who are provided with extensive 'shopping lists'.

Detailed research of the development of the garden has led to a conservation plan being prepared which will see the present 'manicured' area of the garden greatly reduced to approximate its mid-nineteenth century extent. The original driveway will be reinstated as will the estate fencing which was such a feature of the property. Visitors will once again be able to arrive at the front door rather than through a garden entrance. Views to the harbour will again be opened up and gradually appropriate mid-nineteenth century plants will be reintroduced. To do all this will mean relocating the carpark, changing the visitor's entrance, replacing concrete paths with gravel, manufacturing estate fencing and a multitude of other things including convincing neighbours and the public of the reasons for this work. All this is an expensive business. During 1981/82 alone over $100,000 has been spent on new furniture, objects and building restoration and there is still much to be done.

The process of conservation and refurnishing has been made easier since the appointment of a full time curator (Ann Toy) since March 1981. Clive Lucas and James Broadbent provide architectural and curatorial advice respectively and two excellent contract carpenters and a plasterer (who between them manage to do nearly all the necessary building and redecorating) work nearly full time at the house.

The expense of the work is a constant constraint, but while testing one's patience to get on with the works, is probably a good brake, ensuring that there is time to consider the often small but complex issues which arise each week.

Together with the 're-restoration' programme we are preparing new guide books, undertaking intensive training for a voluntary guiding scheme, experimenting with education programmes, rationalising staff duties and cataloguing the collection.

We walk a tight rope between balancing the books whilst maintaining an active works programme. But the results are becoming obvious that with thorough research, conformity to a pre-determined policy, experienced consultants and tradesmen, and daily curatorial care, we can present to the public a property which is as close to its appearance at a previous time as is possible. This is not an end in itself but a means by which we can begin to explain to the public something about the family who occupied the house and the social and political history and taste of their times. (*Summer 1982*)

Above The Dining Room c. 1935.

Right Illustrations such as this one of an unknown Sydney interior thought to date from the 1860s are used to provide information in recreating the interiors. (Mitchell Library)

High country huts

Words and photographs by Klaus Hueneke

Klaus Hueneke has been working on the history of mountain huts and the oral history of the Snowy Mountains for the past eight years. He has recently published a book 'Huts of the High Country' (A.N.U. Press).

To many people Kosciusko means exhilarating downhill skiing, high blizzard-swept mountains, expensive electricity generated by the Snowy Scheme and booming resorts like Perisher, Thredbo and Guthega. It is a place for suave apres ski parties, a simmering dark tan and if you are lucky, another conquest on the bumper to bumper dance floor. Only in getting there from Tumut, Cooma or Corryong does it become apparent that Kosciusko is also a beautiful estate of grand alpine ash forests, cascading cold streams, open frost plains and wind ravaged snow gums. It is so vast in fact that it would take me three weeks to walk from Mt. Nimbo in the north to Mt. Byadbo in the south. Along the way I might startle a feather thatched emu, a grumpy wombat, a scurrying echidna or be screeched at by a Gang Gang cockatoo.

I would also come across the workings of man. Up north it might be Aboriginal bora rings, near Kiandra the water conveying ditches so essential for gold mining, and around mighty Jagungal the fence posts that once carved up the high country into numerous grazing leases. In between, and only if I'm very observant, I might find shelter in one of the many simple rustic huts built over the last 100 years. These lie scattered like pellets from a giant's shot gun in various hidden folds of the landscape.

Some like Currango are fully operational homesteads, others like Pig Gully and Pugilistic are virtually ruins and most like O'Keefes, Broken Dam and Mawsons are small huts with one or two rooms, an open fireplace, some spartan furniture and perhaps a creaking bed or two. The bulk of them were built by graziers and stockmen, a few by gold miners and early ski tourers and in recent years a good number were added by men of the Snowy Mountains Scheme. All are different, all are historic, all are deteriorating and all have at some time sheltered an ill prepared or unsuspecting stockman, bushwalker or ski tourer.

The oldest known structures in the Snowy Mountains are Yans store at Kiandra, the old kitchen and chimney at Gooandra homestead and the Coolamine complex on the Coolamon Plain. Yans store, or what is left of it, has a central portion that was built during the 1860 gold rush. It was occupied for over 80 years, each owner adding another portion or modifying existing rooms. At the turn of the century it had rooms for selling newspapers, meat and groceries as well as a substantial living area. The latter was heated by two big fireplaces, the crumbing remains of which now grace many snap shots and the odd hand painted landscape. One of the chimneys may also have doubled as a baking oven.

Kiandra can be a cold, dreary place and until superseded by Thredbo, Perisher and Spencers Creek, always featured as the coldest place on the mainland. But cold mixed with rain also produces snow and that is a mixed blessing indeed. For the fit and healthy members of the 1860s Kiandra community it meant a pair of speeding butterpats (skis), perhaps an exhilarating jump and much later red hot cheeks in the pub. For the

Cesjacks — a big iron hut with tiny windows, not far from Mt Jagungal.

High country huts

151

sickly and inept however, it meant chilblained toes, waves of goose bumps and a constantly dripping nose. Pink batt insulation, proper orientation to the sun and a passive heat source were unheard of and villagers made do with draughty weatherboard cottages more appropriate for the sunny north. The only structural adaptations were removable gutters so deep snow could slide off easily and stable doors so the bottom half could remain shut.

On the inside, wall lining was either very cheap newspaper or expensive canvas. Sometimes if the landholder was generous or wanted to holiday there himself, the walls would be covered with tongue-and-groove lining board. For research purposes and for discovering the age of a structure, newspaper is by far the most telling. It is also more appropriate to decipher snippets of Boer War news amongst peeling and weathered slabs than to imbibe the same information in the cool and clinical National Library.

At the 100 year old Coolamine homestead I was engrossed for half a day. One room was plastered with 13 layers of the *Albury Banner* and *The Truth*, another was further covered with real floral wallpaper and a third, the small bedroom, was papered with 1940s *Women's Weekly*s. Unfortunately much of this rich patina of many generations had been ripped off and used to light numerous fires in the living room. What had taken early pioneers like the Southwells and the Taylors many hours of sticky fingers and creasing corners was gone in a flash of yellow flame. That, unfortunately, is also the tale of many a mountain hut.

Huts have come and gone ever since Murray of Yarralumla and others first discovered the lush pastures of the high country in the 1830s. Stokes, Tabletop and old Pretty Plain have fallen down with neglect, Windy Creek succumbed to a sliding snow mass, Old Boobee, Alpine and Peppercorn were accidentally burnt and Ibis, Betts Camp and the Kiandra church were deliberately demolished. In many cases the first site chosen had so many good attributes that it was used for the second or third generation hut. Favourable site characteristics included a flat piece of ground, a nearby source of water and dead timber, shelter from westerly winds, an eastern orientation and if possible a view over part of the grazing lease. Hut locations with these attributes include Wheelers, Pretty Plain, Grey Mare, Brooks, Happys, Broken Dam, Four Mile, Mawsons, Whites River, Cascades, Oldfields and O'Keefes.

Wheelers is the shining diamond in these crown jewels of mountain architecture and like all precious stones deserves special care and security. It is a carefully crafted slab hut on the edge of the Toolong diggings on the Khancoban side of the mountains. Built for Wingy (one-armed Will) Wheeler over sixty years ago, its exact age is still a mystery. Lindsay Willis reckoned it had been there a good while in the 1920s, Lila Burzacott told me that it was rebuilt in the 1930s and Errol Scammel showed me a photo of Wheelers at a slightly different site. If the experts with adze and broad axe had dated their work my research would have been so much easier! As things stand and with the application of some historical telepathy I suspect Wheelers was originally built in the 1910s and shifted to its present site in the 1930s.

The most striking aspects of Wheelers hut are the basic layout, the proportions of the east facing main wall and the attention to detail. The slabs are horizontal and neatly fitted between sturdy uprights placed at intervals of one metre. Each upright is precisely morticed into the top and bottom plate and a wooden peg holds the plates to the corner posts. There are five panels of slabs across the end of the hut and seven panels, including doors and windows, along the length. The two windows face east and are symmetrically placed with the entrance door in the middle. A substantial slab verandah runs the full length of both sides of the hut.

The verandahs are full of oddments including a very solid rusty square

Old Currango hut overlooking the Currangorambla Plain and the Brindabellas.

High country huts

water tank, several broken off spades and shovels, an old fence strainer, a chipped grey-blue colander, the bleached hip bone of a cow and a large intact meat safe. This well made safe has a tongue-and-groove wooden floor, a tightly fitting door and mortice and tenon joints all covered with sheet metal gauze. The big iron hooks from which the meat was suspended are still there and with a meat-starved stomach used to dehydrated rations, I could easily visualize T-bone steaks swaying in the evening breeze.

Four Mile hut near Kiandra is a much humbler and smaller structure. It was built for occasional summer use by Bob Hughes the modest bachelor son of the skiing mailman of Kiandra. Bob built his one-man hut in 1937 out of the slabs and building materials left over from the Elaine mine. Most of his old tools, sluicing equipment and reading matter were

High country huts

Left Coolamine — a historic collection of stockyards, iron sheds, log huts and slab homesteads.

Top Oldfields — an iron hut, west of the Brindabellas.

Middle Kellys hut — a substantial weatherboard hut with lath and plaster walls and a brick arch chimney.

Above Vickerys — a log cabin on Journama Creek, near Talbingo, built by Jack Vickery in 1945.

still there when we went to clear it up in 1974. Amongst other things, I was fascinated by a tin of Bisurated Magnesium tablets, a bottle of Blue Ribbon pure vinegar, ski wax made by Victor Sohn in Austria and several novels with foreboding titles like 'Preparation', 'Reconciliation' and 'Enemies' by the Watchtower Society. I was not so impressed by pieces of decomposing gelignite under the bed. It was a relief to jettison them down the nearest unused wombat hole.

Most of the history of the huts comes from the memories of old residents of Kiandra, Tumut, Adaminaby, Jindabyne and Corryong. In the case of Four Mile hut I got many revealing snippets from Tom Yan the bullocky and Jim Pattinson the former ski champion from Kiandra. I also tracked down a large collection of old photographs taken by Bob Hughes with a bellows Kodak camera. A photo of Four Mile in mint condition provided an excellent bench mark for our restoration efforts of 1978 and 1980. The Hughes collection also showed gold mines like the Elaine and Lorna Doone in their heyday.

One of the many surprises in my search for old timers was Bill Hughes, brother of Bob and former resident at the Nine Mile diggings. Bill hadn't been near the mountains for thirty years and with minimal prompting presented a detailed history of the Kiandra goldfields from 1903 (the year he was born) to 1936 (the year he headed north). One of his more dramatic tales concerns Glennie the mountain rouseabout.

Bill Glennie used to have a hut up Four Mile Creek near the old sluicing holes. In old age he lived in Kiandra in a shack on Pollocks Creek where the first gold was found. Young Bill used to visit him especially when Glennie was sick. One day Glennie got a bad foot infection that required rebandaging every day. But the cantankerous old man would tear if off again as soon as Bill had gone out the door. His end came one night when after a heavy dose of whisky he accidentally stumbled into the fireplace. Possibily knocked unconscious and then burnt, a very impressionable young Bill found the half-burnt remains next morning. Life can be rough.

Huts like Four Mile, Broken Dam, Whites River, Mawsons and Cascades are now heavily used by bushwalkers and ski tourers and without the intervention of many volunteer workers would now be in a derelict state. Some of the volunteers started their work back in the 'sixties soon after the last of the stockmen left the mountains. Then in 1971 all concerned groups got together and formed the Kosciusko Huts Association. The Association has had a long and hard struggle convincing the National Parks and Wildlife Service (N.S.W.) of the historic and shelter value of the huts and at one stage was faced with the possibility of losing about half the huts between Kiandra and Kosciusko. But public opinion turned the tide and now over twenty huts are managed as historic sites.

Work on the huts is mainly carried out by clubs, societies and loose-knit groups based in Sydney, Canberra and local towns. A particularly active group are the Bogongs from Canberra. In recent years they have cut, split and adzed new slabs for Wheelers hut, replaced several hundred metres of weatherboard on Broken Dam, rebuilt the chimney of Patons and repaired the windows of Pretty Plain, Witses and Broken Dam. In the process they have collected a variety of adzes, wedges and broad axes, learnt how to use them and applied old and new knowledge with considerable finesse and gusto. It is not unusual to see twenty heavily laden fathers, mothers, singles and others wending their way through many kilometres of snowgum woodland. All tools and materials are carried by hand from the nearest fire trail.

Work parties are a great way of sharing the mountains and of contributing to the presentation of our mountain architecture. The humble huts of Kosciusko are just as important in the national fabric of heritage preservation as the grand old mansions of Sydney or Melbourne or the rock art galleries of Cape York. (*Summer 1982*)

Garden Island

Eric J. Martin

Eric Martin is the Canberra Manager of Philip Cox and Partners Pty. Ltd. who have been the consultants commissioned for the design of the re-use of the three historic buildings on Garden Island discussed in this article. He completed M.B.Env. at U.N.S.W. in 1980 with his graduate project being a building conservation study of the historic buildings on Garden Island.

H.M.A. Naval Dockyard, Garden Island, located on Sydney Harbour's southern shore about 1 km east of the Opera House, is the Royal Australian Navy's main fleet base and principal ship refitting dockyard.

The history of Garden Island dates back to 1788 when it was allocated to H.M.S. Sirius for use as a vegetable garden. Up until the early nineteenth century the island was used as a garden then for the next fifty years it functioned primarily as a pleasure resort for the growing city of Sydney.

During the period of great naval expansion (mid- to late nineteenth century) the idea arose of developing the island as a major naval base. It was conceived in the early 1860s but did not come to fruition until the 1880s.

In 1884 the southern hill was levelled to allow the construction of the naval station. Over the ensuing 12 years such buildings as a Rigging Shed, Barracks, Factories, Workshops, Naval Stores, Offices and Residences were constructed.

The early buildings were in Italianate style with a change to late Victorian and finally Federation style in the later buildings. On handing over the Naval Base in 1896, Rear Admiral Cyprian Bridge described some of

the buildings as being 'almost palatial in character'.

The Naval Base continued to grow and during the years 1940–46 the island was joined to the mainland at Potts Point by the creation of the Captain Cook Graving Dock. Today the fine colonial buildings exist amongst a hotch potch of tin sheds, workshops and dockyard equipment.

Proposals for the modernisation of Garden Island were completed in 1978. The general objective was to modernise the Naval Dockyard and Fleet Base facilities and provide more efficient support for the Fleet. User requirements issued by Defence (Navy Office) in 1977 had the following broad aims.

- To improve the effectiveness and efficiency of the Dockyard and Fleet Base.
- To separate the Dockyard and Fleet Base activities and provide each with the capability of operating independently without duplication.
- To minimize undesirable effects on the environment and improve the aesthetic qualities of Garden Island.
- To retain historic buildings and integrate them with the modernisation proposals.

The Garden Island Modernisation Planning Team established broad envelope forms and development criteria on siting, aesthetics, architecture, materials of construction, finish and landscaping. Consultants from the private sector were commissioned to develop the design for some of the major buildings in accordance with the established criteria.

Part of the early modernisation programme covers the re-use of three of the historic buildings. All three buildings are individually listed and lie within a precinct, mainly the original island, which is entered in the Australian Heritage Commission's Register of the National Estate. These original buildings form a unified group of naval buildings which is extremely rare. Philip Cox and Partners Pty. Ltd. were commissioned to develop the design of these three buildings.

The Rigging Shed and Sail Loft is a handsome utilitarian two-storeyed structure in the Italianate style, featuring arched and segmental open-

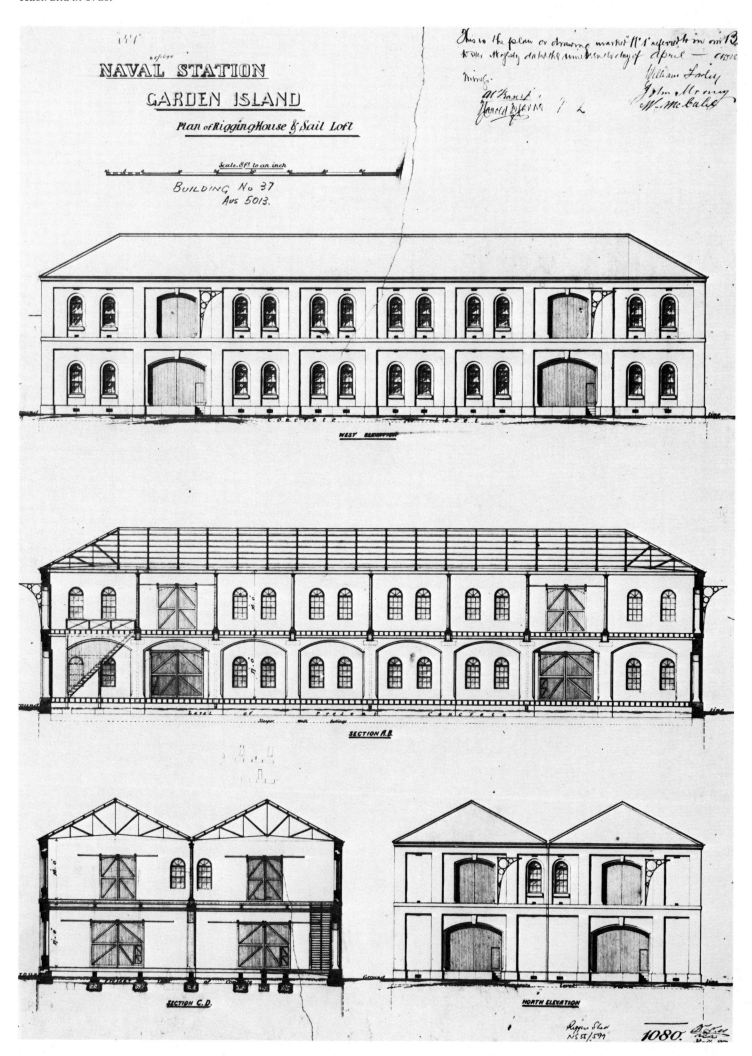

Left Original elevations and sections of Rigging Shed, 1886.

Above Barracks (*left*) and Rigging Shed today.

Above right Naval Stores today.

ings set in recesses in the rendered walling. The original roof was of corrugated iron in a simple hipped form. It is thought that the basic design was prepared in England and documented by the Colonial Architect's Office. The character of the elevations derives from good proportions and the use of repetitive bays.

The exterior and load-bearing walls are of brick with the interior dominated by a central arched brick spine wall with iron girders supporting the first floor. Above the line of the central spine wall, a row of cast iron Tuscan columns divides the first floor and supports a series of lightly framed iron trusses. Originally the entire ground floor was occupied by the Rigging Shed and the first floor by a Sail Loft.

Tenders for the building were called on 19 January 1886, a contract awarded to William Farley on 15 March 1886 and work commenced on site on 31 May 1886. The building was completed in late 1887 for a cost of £16,289.18.11 and finally fitted out by 1889 (at a cost of £623.10.0).

In 1902 the northern end of the Sail Loft was divided off to form a Chapel and relevant windows were amended to allow the introduction of stained glass windows, several in Gothic frames. Internally the Chapel has richly polished timber furnishings, mosaics, memorial plaques and a pulpit in the shape of a ship's bow. A naval gaol and warders' quarters were added to the southern end of the building in 1905 but this was demolished in 1948.

Interestingly, the building was described in the *Sydney Morning Herald* on 11 February 1907 as 'an incongruous and most inharmonious collocation. The one structure embraces a church, a boat store and sail loft, a boardroom, a humble theatre, and a naval prison and warders' quarters.' Today, the former functions survive, for the most part, with a sail loft, rigging shed and chapel coexisting.

The building is significant as it is part of a small group of three historic buildings, is a well proportioned and good example of the light industrial warehouse type of building of the late nineteenth century and contains a finely detailed chapel which is the Australian Navy's oldest church.

Generally, it is intended to restore the exterior of the building utilizing an authentic colour scheme, retaining the rigging shed in the ground floor space, and the first floor chapel, with a second chapel being created alongside it. The remainder of the first floor space is to be converted to an open space to accommodate office staff as part of an adjacent Refit

Australia in Trust

Berth Support Facility.

These measures are intended to enhance the building's cultural significance. Fortunately, original working drawings and early photographs are available to form the basis of the proposed work.

In detail the work includes removal of external stairs and the introduction of internal fire isolated stairs in a manner not to jeopardise the integrity of the building. They will not interfere with the external fabric. Removal of redundant service lines, signs and boxes, upgrading of existing amenities and removal of first floor mezzanine are also included.

The single-storey Kitchen and Service Block to Barracks was linked to the tall barracks block by a covered way. The barracks follow a typical British pattern, with three levels of Tuscan columned verandahs to the west elevation and a completely symmetrical arrangement.

The kitchen and service block originally housed kitchen, wash house, lavatories and baths around a central hallway. Documentation of the work was undertaken by the New South Wales Colonial Architect's Office under the jurisdiction of James Barnet as were all the early buildings on Garden Island.

Walls are rendered brick with footings and floors of mass concrete. The simple hipped roof is slate. The original fenestration of 12 paned double-hung windows has been modified somewhat.

It was constructed together with the barracks; tenders were called on 30 November 1886, contract let to G. Langley on 7 January 1887 and the entire job completed in 1888.

The Barracks building is a classic example of colonial architecture constructed during the second half of the nineteenth century. Generally, the facilities at present accommodated are signwriters' workshops and amenities and a duty beat driver's rest room. Internally the building has been amended to accommodate these users.

The intended changes are restoration of exterior, especially the fenstration, restoration of covered way linking to Barracks and central hallway and openings. The building will be converted to accommodate cool room, provision room and clothing store.

These proposals, as for the Rigging Shed, are intended to enhance the building and its significant contribution to the historic setting. The basis

Bottom Original working drawings of Barracks, 1886.

Below Interior of Naval Stores.

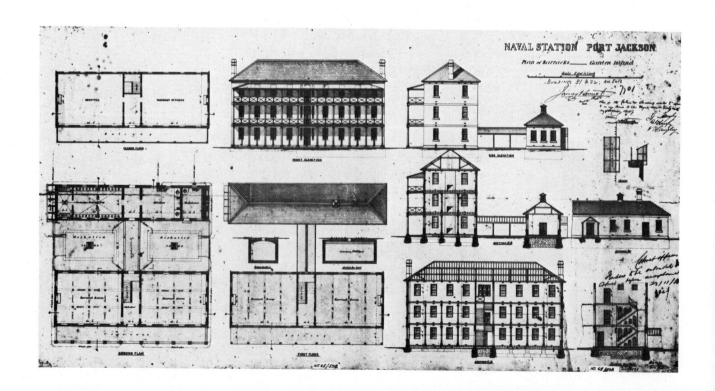

of the work lies in the original working drawings. The proposed new uses will generally allow the fenestration to be restored. It is envisaged that the facilities required for the stores and cool rooms will not interfere with the external fabric or the quality of the spaces.

The Naval Stores is a four-storey (semi basement and three upper floors) late Victorian warehouse of load bearing polychrome bricks with sandstone string course, cornices, sills, copings and granite thresholds to doors.

The building is divided into five fire separated compartments by vertical cross walls. Within each section are two rows of circular cast iron columns supporting iron girders, timber joists and a 50 mm tallow-wood floor. The ground floor was paved with Val-de-Travers asphalt, 38 mm thick. Lightly framed wrought iron roof trusses span between the masonry walls and support the roof. The roof was originally corrugated galvanized iron.

To service the stores, five externally mounted wrought iron whips are provided above the large arched doorways on the northern side. Between these doorways and around the rest of the building are numerous windows of varying size. The timber box framed double hung windows have two and four light sashes. In the centre of the northern and southern parapet, carved in sandstone, is the Royal Cypher of Queen Victoria and the date 1893. The original water operated hydraulic plant consisted of engines, accumulator, five hoists (whips) and two lifts.

Tenders for the foundations were called and contract let in 1891 but tenders for the main building were not called until 1892. The contract was first let to J. C. Waine, but it appears that the 1890s depression had its effect as they did not proceed with it and the contract was awarded to Howie Brothers shortly afterwards. The building was completed in early 1894 (even though the date on the building was 1893) at a cost of £31,886.

This building is a particularly fine example of a Victorian warehouse building which also exhibits a strong naval character. It is one of the few Victorian warehouses remaining in Sydney and provides an interesting contrast to the more classic earlier buildings on the Island.

Due to the increased demand for office space it is proposed to convert this and an adjacent more recent building (constructed in 1939) to an Office Complex. The Naval Stores will be relocated away from Garden Island.

The conversion will include the removal of all partitions, stairs, etc., from the four floors to provide five clear bays broken only by the original cast iron columns and brick fire walls. Service cores containing lifts, firestairs, toilets and plant rooms will be constructed on each level. The timber floor structure will be topped with 50 mm concrete and carpeted in offices. Internal walls will be painted except in wet areas where walls and floors will be tiled. Generally, the underside of the existing timber floor structure will be exposed to form the finished ceiling in office areas.

The adjacent 1939 building will be upgraded and converted to staff amenities and services on the lower floors, and offices on the upper floors. The external fabric of the building will be restored including removal of redundant service lines, signs, assorted boxes and poles, cleaning of the building, repairs to stonework and brickwork.

Garden Island is an important part of Australia, not only because it is part of Australia's national heritage but also because it is a modern and important naval base. It is a significant place in our National Estate because most of the original naval station buildings constructed between 1885 and 1896 remain and form an unusually unified group.

The development that has just begun as part of a total and ongoing modernisation illustrates that existing building stock can be effectively used to meet modern requirements yet not compromise the significance of our heritage if carefully and sympathetically handled. (*Winter 1983*)

Carpenter's decoration

Kate Blackmore

Kate Blackmore has recently completed her thesis on Detail of small scale Sydney architecture, c.1850–c.1890: a survey of sources, *for the degree of M.Sc. (Architecture) (Conservation) at Sydney University.*

When Joseph Fowles described Sydney in 1848 in his work of that name,[1] he was describing a simplicity, an orderliness that in architectural terms could best be described as Georgian. This of course does not mean that Sydney consisted of row upon row of symmetrical boxes with hipped roofs, multi-paned windows and shuttered french doors. The horizon of Sydney Cove, like many rural landscapes, by this date was dotted with castellated turrets and paired gables. Public buildings bowed to Classical form and churches emulated their medieval forebears. But Fowles' Sydney was a city in its infancy in both size and style.

By 1900, only fifty years later, an amazing metamorphosis had taken place. Multi-storeyed buildings had now created a new skyline. The passionate excesses of mid-Victorian decoration were immortalised in commercial building facades of the 1860s and 1870s. Sydney sported its slums — areas of sub-standard housing and hygiene. But perhaps the most dramatic change was the sheer growth of suburban Sydney and thus the proliferation of the suburban home.

It is these houses, alternatively the 'renovator's special' or the 'classified residence' that are now, arguably, the clearest statements of various periods in late nineteenth century Australian architectural history.

Clearly, a great majority of these houses were not the individually designed creations of the architectural practices of the day but rather the result of enterprising speculative building and development — often along the ever-extending suburban railway network.

Close to the city, where population density was greatest, the terrace house was the logical building type to obtain maximum use of available land. As Victorian Sydneysiders moved out to the suburbs, where land was not at such a premium, the free-standing house became more typical. Between 1850 and 1890 thousands of homes were constructed for a Sydney population which by 1891 had reached nearly 400,000.[2]

The number of building styles which appear in this period defy individual description. The terminology used to discuss architectural styles of Victorian Sydney (and elsewhere) is not strictly concerned with the building styles of these small houses — one would be hard pressed to find a truly Italianate cottage. What can be found, however, is the decorative elements or ornaments of particular nineteenth century styles applied to such humble constructions. Thus a distinction may be drawn between style of building and style of ornament.

In Sydney the years 1860–80 saw the proliferation of four basic building styles: those which followed the Colonial Georgian tradition; those with sufficient elements in the picturesque Gothic tradition to be labelled Victorian Gothic; the simple 'L' plan house with its decorated gable end and odd mixture of Italianate, Gothic and Victorian terrace ornament (for convenience of description here termed 'Italic'); and the ubiquitous terrace. The style of ornament used on these houses ranged from the plaster (later cement) follies of Victorian Italianate through the delightful timber 'dressings' of Victorian Gothic and the more sombre detail of Colonial Georgian to elements without a specific style such as cast iron.

Of the building styles outlined above, certain ones lent themselves more than others to carpenter's decoration. The Victorian terrace resorted more to the plasterer and to cast iron work for its ornament. The 'Italic' house used a combination of styles and materials for its decoration. But the Victorian Gothic and Colonial Georgian styled houses were the happy recipients of many decorative timber elements some of which, given minor modifications, retained their currency into the twentieth century.

Buildings constructed in the Colonial Georgian tradition exhibited a number of characteristic timber details. French doors and shutters and six-pane sash double hung windows head the list together with the six

Victorian Gothic ornament. (Photo: A.H.C.)

House in the Georgian tradition with typical early and mid Victorian timber detail. (Photo: A.H.C.)

panel front door. This latter feature had, by the 1850s, been replaced by the four panel door — a style of door which was to retain its popularity for many decades to follow. As late as 1890, timber merchants were still advertising six-pane sash windows — proof of the continuing life of this feature.

Other characteristic timber features of these houses were balustrading to first floor balconies (when built in two storeys) and verandah details of simple linear character. By the 1870s, however, cast iron had practically usurped the role of timber in use on balconies and awnings.

The progress of the picket fence, from its elementary beginnings through a delightful array of 'dressings', appears to have been unaffected by the march past of building styles. By 1890 one Sydney mill could advertise nearly thirty designs in dressed or turned pickets.[3] The earlier examples, with flat or 'spade' topped pickets, usually had a flat profile in elevation. The scalloped profile, which appears to have commenced its life in conjunction with the Victorian Gothic style, was later used at random on all manner of styles of the late nineteenth century.

The timber details of Victorian Gothic are, by their nature, more immediately apparent than those of the Georgian derived building. Two features which had their Australian architectural origins in this style but were retained for use in later styles — including the 'Italic' style mentioned earlier — were the timber barge board and finial. A scalloped edge was common to barge board designs of all dates. Other patterns commonly traced in the timber were more obviously Gothic in inspiration — in particular the trefoil and quatrefoil. The turned finial which terminated the gable end was invariably of simple pattern and essentially unchanged throughout this period.

Other timber details common to early and mid-Victorian houses were the use of lattice — a fashion destined to reappear again and again — and decorative timber brackets for verandah posts. The fretwork brackets of Victorian Gothic houses were often 'extended' horizontally to form a valance. Often the valance created a colonnade effect along the verandah.

Carpenter's decoration

Typical Georgian style house with Victorian timber detail. (Photo: A.H.C.)

The simple weatherboard cottages which mushroomed throughout New South Wales around the turn of the century seemed to reject iron as inappropriate and revived these timber ornaments.

Less frequent by this period was the use of timber for classically derived ornaments such as the 'acanthus' or 'palmette' decorations so typically rendered in cast plaster on Victorian terraces or shopfronts, or the Greek key motif frequently inscribed on the pilasters of quasi-Regency buildings of the 1850s.

Windows, in all manner of styles, attracted a variety of timber decorations throughout the latter half of the nineteenth century. One feature which dates in New South Wales from at least the 1840s — and perhaps earlier — was the window pelmet. The pelmet was designed initially to house timber venetian blinds (externally) which provided shade and privacy to the first floor french doors of city townhouses. By the 1880s it was often a non-functional decorative detail on the windows of cottages.

Dormer windows on houses decorated in the Victorian Gothic tradition were treated to the same timber ornament as the gable end barge board and finial. A new development late in this period was the window hood. The typical window hood of the 1880s combined the overhanging eave of the gabled dormer window and the fretwork 'fringe' of the pelmet with the hipped roof of a bay window — so characteristic of Victorian Gothic houses. The end result, a suitably functional device for the Sydney climate, became the prototype for a variety of twentieth century aluminium and canvas accretions.

As with the external details mentioned above, certain internal details lent themselves more readily to carpenter's decoration and are common elements to many small houses. If an overall trend may be discerned it is that while craftmanship may have maintained technical quality, aesthetically many timber details suffered from a coarseness and vulgarity seen throughout the decorative arts of the late Victorian period.

Stair details tended to become more ornate and bulky over this period. The profoundly simple newel post of an 1850 house was akin to the

Far left Window pelmet, c. 1880.
Left Late Victorian window hood.
Below Victorian Gothic ornament. (Photo: D. Liddle)

baluster of an 1880 staircase. By this latter date firms such as Goodlet and Smith's Victorian Steam Saw and Moulding Mills were producing countless designs for the various stair members catering for all tastes in true Victorian style.

Similarly, the timber mantelpiece — at least those designed for use in principal or public rooms — became more lavishly decorated as the century progressed. Earlier examples tended to bear simple geometric designs or paired 'acanthus' leaf decoration. By 1890 the entire face of the mantelpiece could be a riot of Italianate motifs.

Timber mouldings, whether for cornices, skirtings or glazing bars, suffered from the same coarsening noted above. However, on the scale of house discussed here, it was unusual to find extravagance. Time honoured profiles, with their origins in antiquity, were used repeatedly throughout this period and for many decades to follow.

Obviously there were many more decorative devices wrought in timber than those few briefly outlined here. There was, for instance, the somewhat unusual practice of using weatherboarding to imitate ashlar or the fleeting appearance of triangular timber pediments to terraced houses about 1880 — in fact a revival of an 1850s practice. There were other common decorative timber elements such as gates, posts, columns, brackets and consoles, shutters and countless other details which went towards characterising particular periods in late nineteenth century building practice.

In the late 1880s brick made a startling reappearance in Sydney. Simultaneously grand houses in the style termed Queen Anne and later, Federation, sprinkled the suburbs. These two influences combined with the Depression of the 1890s and the commencement of an aesthetic trend away from the excesses of Victorian taste resulted in an overall simplification in the detail of small houses. While decorative elements of the Victorian period reappeared in the first decades of the twentieth century there was an austerity of design and simplicity of style which was to have a lasting impact on suburban housing and sound the death knell for many of the picturesque details examined here. (*Summer 1983*)

First Government House, Sydney

Helen Proudfoot

In 1983, Urban Historian Helen Proudfoot completed an historical study on the First Government House, Sydney, and the subsequent history of the site after 1845 for the Premier's Department and the Department of Environment and Planning.

On the fifteenth of May, 1788, nearly four months after landfall, Governor Phillip laid an inscribed copper plate marking the foundation of his Government House. The inscription was done in a fine copperplate hand:

His Excellency Arthur Phillip, Esq., Governor in Chief and Captain General, in and over the Territory of New South Wales, &c, &c, &c, landed in this Cove with the first settlers of this Country, the 24th Day of January; 1788, and on the 15th Day of May, in the same year, being the 28th of the Reign of His present Majesty, George the Third, the first of these stones was laid.[1]

On 20 February, 1983, nearly two centuries later, the foundations of Phillip's building were re-discovered in the first weeks of the archaeological investigation of the site now being carried out under Anne Bickford. This is surely the most exciting archaeological find of European civilization in this country. Astonishingly enough, though the building had been demolished in 1845, the site, though located in the centre of Sydney, on the corner of Bridge and Phillip Streets, has escaped large-scale redevelopment up to the present time, being used as a carter's yard for some decades during Victorian times and then for a two-storey corrugated iron Department of Works building from 1912 to 1970, which did not cover the whole area of the site. Since then it has been used as a government car park.

Suddenly with this archaeological investigation we are transported back to 1788, the first year of white settlement. The stratigraphy of the site, in its subtly different layers, shows the history of development: construction, extension, further extension, demolition, adjacent road building, hiatus, subsequent building and demolition, and levelling over and sealing off as a car park. As the trowel peels off the various layers, we find, only a few metres below the surface, the stones hewn by the first convict settlers, and the bricks made by their hands. This building was the first major permanent structure erected in Australia; its bricks are the earliest evidence of building fabrication we have; this was the very spot from which Phillip directed the affairs of the thousand people who had arrived on the First Fleet; the building became immediately the focus of the town of Sydney.

The very fact Phillip decided that, despite the 'want of lime',[2] he would go to two storeys, indicates that it was intended not just as a dwelling for himself, but as the principal building of the settlement, a symbol of stability and permanence. 'The House intended for myself was to have consisted of only three rooms', he wrote to Lord Sydney, 'but having a good foundation, has been enlarged, containing six rooms, and is so well built that I presume it will stand for a great many years'.[3]

The building was also modestly embellished with a pedimented breakfront, giving emphasis to a front door graced by a semi-circular fanlight

and side-lights. There was a blind roundel within the pediment, both at the front of the house and at the rear stair-hall. The building possessed the Colony's only staircase, a source of wonderment to Phillip's Aboriginal visitors, who were astonished to find people walking about above their heads.[4]

The designer of the building may have been Henry Brewer, a naval man who had known Phillip for some years before he accompanied him to New South Wales. Brewer was a skilled draughtsman and was active as a building superintendent during the first two years when all efforts were concentrated on building activities in the Colony. Phillip himself may also have had a hand in the design. James Bloodworth, we know, was the superintendent of the convict bricklayers who worked on the building.

This first major administrative building was built in the vernacular English tradition of the Georgian era. It had a dressed stone base, but the walls were built of brick,[5] rendered with pipe-clay which was also used as a mild binding agent for the mortar.[6] Clay tiles were burnt early, and may have been tried, but the roof appears to have been clad in she-oak shingles, coloured grey or blue-grey on the early drawings.

It is in fact the early pictures of the settlement which have provided the most useful evidence of the building and its construction and setting. There are comments about it by Phillip and the early diarists, especially Tench, White, Collins and Worgan, and the odd passing remark by King and Mrs. Parker, a visiting sea-captain's wife and other visitors, but the building is not described in great detail, and no original plan has survived. The only early plan we have is that sent in a letter by John Palmer and countersigned by Governor Bligh when ordering 'Floor Cloths' for the rooms for the year 1807,[7] and this shows the front rooms only of Phillip's building and the long Drawing Room added by Governor King in 1800.

A more elaborate plan of the whole complex as it was extended several times by successive governors was made by Mortimer Lewis in 1845 when he reported on the building.[8] By that time, of course, it had grown considerably from its original 55 feet or 18 metres, and was over 50 metres in length. This plan shows these extensions and the range of outbuildings at right angles to the main structure — kitchen, laundry, bakehouse, privies and possibly servants' rooms — all necessary to the func-

First Government House, Sydney

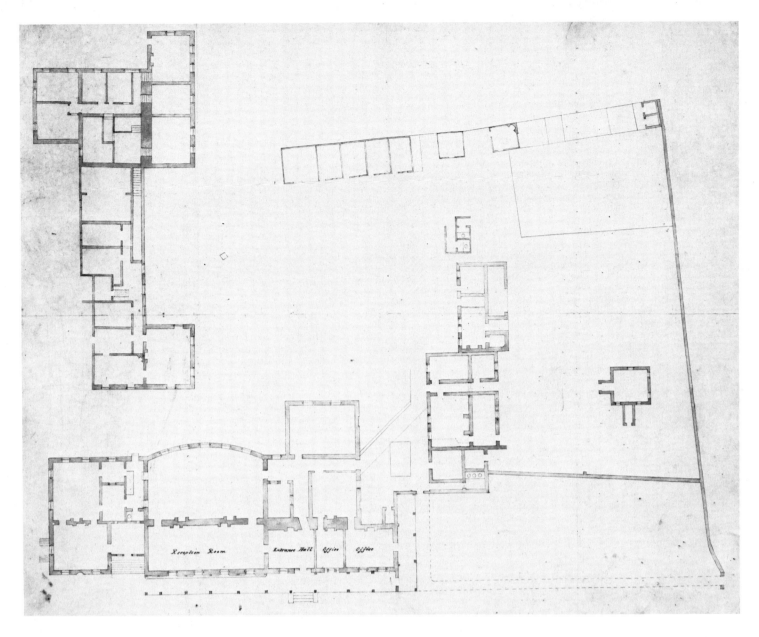

Left William Bradley. Governor's House, Port Jackson, 1791. From his voyage to New South Wales, Sydney, 1969. (Mitchell Library)

Above Floor plan of Government House, 1845. The plan accompanied the *Report on the present state of the Old Government House* prepared by Mortimer Lewis on 15 Sept. 1845 for a Board of Survey appointed to examine the state of the building. (A.O.N.S.W. Colonial Architect Papers 4/2717.2) The plan at the top left corner of the drawing is that of the first floor, turned at right angles and does not represent any building on that side of the site.

tioning of the house and an integral part of the complex, helping us to understand how life went on in the early days of settlement. There were also two cellars underneath the two western front rooms.

The accuracy of Mortimer Lewis' plan is at present being tested on the ground by the archaeological investigation. The footings of the back wall of Phillip's building have been found, as well as the returning wall of the rear stair-hall, and the western wall. Brick footings from various walls in the kitchen wing and an elaborate drainage system have been uncovered.

The contrast with the surrounding development could not be greater. This part of the City of Sydney is now very intensively developed with the huge A.M.P. building opposite, and a hotel tower rising behind the old Treasury building. Circular Quay was long ago reclaimed and re-designed and the shore-line pushed further out. The Tank Stream, once flowing down to a sandy shore is buried beneath Pitt Street.

The early colonial pictures of Sydney Cove flesh out the documentary evidence of the building. Government House stands out as the main building amongst the little group at the Cove. Its rising walls are shown both by William Bradley in his view *Sydney Cove, Port Jackson, 1788* and by Captain John Hunter in his *View of the Settlement on Sydney Cove, Port Jackson, 20th August, 1788*, made into an etching by E. Dayes in London and published in Hunter's account of the Colony.

Bradley also drew the completed house in 1791, with a kitchen wing, a number of small out-buildings, and a sentry box on each side. A most interesting collection of early drawings and water colour sketches, is in the so-called 'Watling Collection' in the British Museum (Natural History) in London. Several of these are by Thomas Watling, the 'Limner of Dumphries' who was transported as a convict. One is by Charles Raper and another by an unknown painter. They show the house and the Cove from various aspects, from the north, west and south and they immediately give us a realistic impression of the building itself and of the Colony as it must have been in the 1790s. The best known of Watling's early pictures however, a fine oil painting entitled *A Direct North View of Sydney Cove in 1794*, is fortunately held in the Dixson Galleries, Sydney. This shows the house with a squared, cultivated garden running down to the Governor's Wharf at the Cove in front, the storehouse and officials' houses between it and the Tank Stream, and rows of tiny convict huts behind.

A later drawing by John Hunter (*c.* 1893) in the La Trobe Library and an etching by W. S. Blake, show the addition of the front verandah; William Westall's view (*c.* 1800) in the National Library, Canberra, shows Governor King's extension, without verandah, which was added shortly afterwards. W. L. Lewin, William Evans, John Eyre and J. W. Lancashire all depicted the house during the King and Bligh periods. The only picture located showing the interior of the house is a contemporary cartoon of Governor Bligh's arrest in 1808, showing him being dragged from beneath a bed in an upstairs room by a red-coated soldier. Bligh was then virtually imprisoned in Government House for about twelve months.

Governor Macquarie's first extension was in 1811, when the bricklayers under Thomas Legge constructed a large Dining Room or 'Saloon' with a shallow five-windowed bow and rebuilt the rear of Phillip's building. He had another go in 1819, with the help of Francis Greenway, adding a gabled eastern wing of additional bedrooms and so completing the northern facade of the building to form an asymmetrical rather Italianate composition overlooking the picturesque Sydney Cove.

Sophia Campbell in her drawing of *Sydney Cove in all its glory, 1817*, C. Cartwright in his plan of the *Government Demesne*, and an unknown engraver, have all shown the building after Macquarie's first extensions were done. For the second stage, we can turn to paintings by Richard Read, Snr., and his son, also Richard Read, and to the sketch done by G. T. W. B. Boyes in 1824. It is just as well, as Macquarie was careful not to mention his extensions in despatches after being cautioned by Earl Bathurst about the extravagance of his building programme, so we do not have an official description of them.

When Governor Darling added his part to the rear of the building, on the other hand, we have the official description but no pictures of the rear, apart from the Lewis plan. Augustus Earle did a watercolour and lithograph in 1830, but the clearest representation of the building's later phase is the watercolour by Charles Rodious in the Mitchell Library.

William Westall. Government House, Sydney, 1802 (1800?), watercolour. (Original in Mitchell Library)

Conrad Martens also painted it in oils in the 1830s and then G. E. Peacock painted it as it was just as Sir George Gipps vacated it in 1845.

So we have to depend a great deal on the information we can glean from these contemporary pictures. The archaeological investigation will be able to tell us more about the materials that the building was constructed of and the way of life of the governors.

The building was more than simply a house. It was as well the administrative centre of the Colony, and the main meeting place where the Governor consulted his officials, received the colonists, made his decisions for the regulation of the Colony, and wrote his despatches back to England. It was in the centre of the activities of the town, close to the port and busy George Street. Sentries paced up and down in front of the house, carriages rolled up to the front door, visitors waited in the hall, on festive occasions the military band played outside. There was a promenade along the waterfront from Macquarie's time, a favourite walk for the citizens of the town. The *Sydney Gazette* was published from one of the rooms of the complex; the Legislative Council, newly formed in 1824, held its first meetings in the Drawing Room. Governor Macquarie invited the emancipists to his table, thus signalling a new phase of the Colony's development. Governor Darling entertained assiduously making the house the social centre of the community. Children were born there and some died there; explorers were farewelled and feted; foreign visitors were hospitably received. In all, there was a complex range of activities conducted on this site, activities which were of vital importance to the growth and development of the Colony for over 50 years.

Just as the building itself, in its expansion, reflected the growth of the Colony and its activities, so its grounds changed over the period also. At first there were carefully cultivated and carefully guarded rows of vegetables in squared garden beds in front of the house running down to the water's edge. Watling shows these in detail in his painting of 1793. Then, during Governor King's time a few small trees appear dotted along the squared beds, possibly fruit trees, with ornamental pines to the west of the building and other trees near the water's edge. The pines which actually survived the house for some decades in Bridge Street were probably planted at this time. Once the lean years at the beginning of the

Richard Read, snr. View of Government domain..., c. 1819, watercolour. (Original in National Library of Australia)

settlement were over, the garden became less utilitarian and more ornamental. By 1807 in fact, Governor Bligh had decided to sweep away all the old cultivated allotments, and had it laid out 'in Walks with clumps of trees',[9] smoothing out the rocky outcrops, to make a 'Pleasure Ground' in the fashionable English landscape style. The paintings of Read, Earle, Martens and Peacock all show this later phase.

Government House was demolished in 1845, on the recommendation of Mortimer Lewis, to allow Phillip Street to be continued on down to the Quay, and Bridge Street to extend to the new Government House gates in Macquarie Street. The intention was to sell off the allotments on the eastern side of Sydney Cove to raise money to pay for the building of the new Government House. The site was offered to the City Council for a Town Hall, and accepted, but at the very last moment the Council decided not to go ahead. The streets were laid out, the site was subdivided, Raphael Place was formed, and the peripheral terraces along Phillip Street and Young Street were built during the 1870s and 1880s.

The site, then bounded by the surveyed Bridge and Phillip Streets, was used as a drayman's yard during the latter half of the nineteenth century, and was surrounded by wooden hoardings. There were a couple of small shops at the footpath, and a small wooden house within the yard. In 1912 the Department of Public Works constructed a two-storey corrugated iron building on the site which lasted until 1970. It then became a car park.

Now, with the proposal to build a 38-storey building on the site by Hong Kong Land and the State Government Board, it seems unlikely that the foundations of Phillip's Government House and its out-buildings will be able to be retained intact, unless there are major modifications to the planned development. The Minister for Planning and Environment has not placed a Conservation Order on the site and so these most historic relics have no legal protection.

We Australians take a lot for granted. We enjoy the life that our forefathers made possible for us, we inherit their culture and their way of life. But we do not honour them. Instead of these archaeological findings being hailed with delight they are seen by some as a possible impediment to more foreign investment. (*Summer 1983*)

Port Arthur Historic Site

Brian J. Egloff

Dr. Brian Egloff is the Project Manager of the Port Arthur Conservation and Development Project. In the past few years his major concerns have been with cultural resource management projects in the U.S.A., Papua and New Guinea and Australia.

In the past events have occurred which either by their scale or intensity become enshrined in history as larger than life. Port Arthur has in the past and is to some extent today the scene of just such events. As the centre-piece of an Imperial penal complex, Port Arthur with its satellite stations on the Tasman Peninsula, enjoys a notoriety inextricably linked with the evils of transportation and the dark days of Van Diemen's Land. So strong was the taint of the convict past that following the closure of the prison in 1877, the townspeople who came to reoccupy the abandoned residential buildings chose to name their community Carnarvon.[1] This masquerade lasted until the ever increasing lure of the tourist's shilling brought about the partial reassertion of Port Arthur's place in history — so vividly portrayed in the 1920s epic film *For the term of his natural life*. Today Port Arthur is the scene of an ambitious and in many respects unique conservation and development programme.

Readers of *Heritage Australia*, as the converted, know that conservation is fraught with qualitative as well as technical concerns. Port Arthur is an extreme case in that for 50 years evidence of its previous penal status was systematically destroyed, and that which remained was under continual threat.

In 1913 M. Bucirde reported:

But taking into account the general aspect and condition of Penitentiary, I should recommend that the buildings be pulled down and I consider that the magnificent view of the bay to be obtained from the high road above the Penitentiary when buildings are down would more than compensate for the loss of same from a tourist point of view.[2]

Attitudes seldom change quickly, but certainly actions of the nature proposed by Bucirde of the Tasmanian Public Works Department in 1913 are not widely exposed today, or are they? After all it was only relatively recently that the convict parade ground in the very heart of the historic settlement was used as a caravan park with the adjacent guard quarters being converted into a wash house. Today the visitor to Port Arthur is confronted with perplexing juxtapositions of old features and new elements which continually highlight the scenic recreational qualities of the place at the expense of the very historic values which presumably draw people to the site. Few people realise that an historically significant convict built slipway was filled to provide parking space for the commercial/recreational wharf at Port Arthur.[3] While three kilometres of public road ensure 'drive-through' scenic viewing, the historic site is cut into disjointed unrelated fragments. A proliferation of informal picnic spots with 'immediate' vehicle access again brings home the impression that Port Arthur is by and large regarded as a generalised recreational facility offering a plethora of attractions.[4] Not the least of these being a cricket pitch, powerboat moorings, bus tours, gas barbecues, motor launch tours, boat ramp, on-site parking, motel accommodation overlooking the bay,

public bar, a youth hostel, a souvenir shop and cafe built upon the site of convict overseers' quarters as well as sufficient mown grass to support numerous informal footy games. All this lies within what was once the most notorious penal institution in Van Diemen's Land.

Quite obviously Port Arthur would not have been developed in such a fashion over the past half century unless there were such demands. Now that the National Parks and Wildlife Service is rationalising the conflicting demands it is apparent that the enshrined use of the historic site as a generalised recreational area will not be easily put aside. The advent of a substantial joint Federal/State grant has enabled the Tasmanian National Parks and Wildlife Service to attempt a redirection where greater emphasis is placed on the historic qualities of the site. How much of the continuing demand to use the site as a genteel manicured picnicking spot comes from a lingering reluctance to face the tainted convict past, what the historian Peter Bolger refers to in *Hobart Town* as the 'convict cringe'? Perhaps it is true that many Tasmanians would prefer to use this historic site as a playground rather than come to grips with their convict past. It is obvious that the conservation and development of Port Arthur as an historic site is far more than the basic processes of fabric conservation or restoration; it is an exercise which demands a fundamental change involving no less than the search for a plan which will respect the historic qualities of this most significant and evocative site.

In 1830 Port Arthur entered the colonial scheme as an unambitious timbering station selected in part due to the ready supply of timber and a sheltered harbour as well as for the reasonably secure peninsula location in close proximity to Hobart.[5] Shortly thereafter the penal settlements at Macquarie Harbour and at Maria Island were closed, and their inmates transferred to Port Arthur, as at a later date were the prisoners on Norfolk Island. The settlement at Port Arthur grew as a markedly varied agglomeration of brick, stone and weatherboard buildings which on the whole were used not only to shelter or incarcerate the prisoners but to provide them with appropriate activities. At one time the settlement was a major industrial complex out of which flowed prodigious quantities of sawn timber, split shingles, forged nails, clay bricks and shoes. During the early years of the settlement the shipyard at Port Arthur constructed numerous vessels.[6] What is now known as the 'penitentiary' was one of the largest grain mills in Australia until its conversion about the mid-point of the nineteenth century.

Port Arthur developed as an industrial centre with a network of supporting satellite stations, all of which were dependent upon convict labour. As such that labour either had to be trained or upon arrival possess sufficient trade skills. The neighbouring 'Boys' Prison' at Point Puer developed a successful training programme in the manual arts, placing through a depot in New Town its apprentices in positions outside of the penal system.[7] Nevertheless there was always a shortage of tradesmen represented in the convicts and skilled overseers proved difficult to recruit particularly during the years of the Victorian gold discoveries. Port Arthur produced as much as practically possible the materials which were used in its building programme, except for exotic materials such as raw iron, lead, glass and some finished metal wares.

Developments at the outstations on the Tasman Peninsula were equally dramatic, including a complex coal mine,[8] a windmill,[9] as well as the convict pushed railway which transported goods and small boats from one coast of the peninsula to the other. Relatively cheap convict manpower was the key to this development. When that labour was not forthcoming as transportation to Van Diemen's Land ceased in 1853 and the Imperial subsidy diminished, the system ground to a halt.[10] Port Arthur gradually changed from a vibrant industrial complex to a moribund sta-

Port Arthur Historic Site. (Photo: Max Dupain)

tion for the recalcitrant and enfeebled prisoner being housed, fed or cared for in the Hospital, Asylum and Paupers' Quarters. A rather ignoble close to a grandiose Imperial venture which came to be a Colonial liability.

Following the close of the station in 1877 it proved difficult to foster alternative uses for the facilities which then gradually fell into disrepair, were salvaged for building materials or fell victim to the ravages of bush fires. Victorian curiosity seekers sought Port Arthur, thus establishing a tourist image for the site as well as ensuring that those buildings servicing the trade remained in relatively good order. There are veiled rumours that arson contributed to the removal of those penal structures which did not fit the image of a refined Victorian touring resort.[11] Shortly after the turn of the century the State of Tasmania began to reacquire Port Arthur, vesting ownership in the Scenery Preservation Board, where it remained until 1971 when the National Parks and Wildlife Service was established.

This brief history of Port Arthur only touches upon the more salient

The Port Arthur conservation and development project

National Parks and Wildlife Service, Tasmania
P. Murrell, Director
Project cost: $9,000,000
Duration: 79/80 to 85/86 (7 years)
Objective: to conserve and develop the historic qualities of Port Arthur and the region

Particular emphasis is placed upon the Port Arthur Historic Site with a programme designed to:

■ establish the significance of the Site's various features respecting all periods of Port Arthur's history;

■ prepare and initiate a comprehensive Site Management Plan;

■ research the historic and social qualities of the Site, and perform appropriate recording activities;

■ develop the Site infrastructure to meet with conservation and visitor requirements;

■ conserve and stabilize the major standing ruins;

■ restore and adapt intact structures to a condition which will ensure their future conservation by re-establishing a pattern of use;

■ restore the visual integrity of Port Arthur as an historic site by resiting or masking intrusive elements;

■ establish an interpretation programme for the Site which will imaginatively develop research findings into an enriching experience for visitors;

■ develop facilities for visitors to Port Arthur;

■ develop a general conservation scheme for Tasman Peninsula.

The programme is being carried out under the guidelines of the ICOMOS (Burra Charter) with an on-site staff which includes: manager, archaeologists (2), architect, curator, draftsperson, extant recorder, guides (2), historian, interpretation officer, liaison officer, park asst. (7), secretary, senior park asst. (6). The project staff works with the NPWS site management component, made up of 21 individuals.

Port Arthur Historic Site

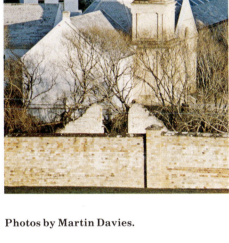

Photos by Martin Davies.

aspects of the site's history and to some extent leads towards an understanding of the equally complex problem of conserving and developing this unique historic site. Paramount in this equation is the establishment of significance for the whole and each individual part as well as the explicit definition of priorities. On a somewhat smaller scale this involves single buildings which may have served a number of significant functions during their days, which speak for a specific role such as a residence during the penal period or as a community facility during the later Carnarvon period. As such the Commandant's residence became the Hotel Carnarvon, the Asylum became the Carnarvon Town Chambers, the Church of England parsonage became the Carnarvon Post Office, the Senior Medical Officer's residence became the guest house 'Clougha'.

The Australia ICOMOS Charter for the Conservation of Places of Cultural Significance (The Burra Charter) that guides the Project states that each significant historic period must be respected. As such the Victorian tourism theme of the site's history will be featured side by side with convictism. Strictly speaking conservation should result in as few noticeable changes as possible. As such the conservation of the major ruins at Port Arthur, the Hospital, the Penitentiary and the Church is being designed to stabilize the fabric.[12] This in turn will permit closer access by the public. Convict built officers' cottages will be restored and either presented to the public as display houses or will serve as Historic Site management facilities. Port Arthur is 'the real thing' and as such care is being taken to retain that essential patina of history.

At all times the conservation of the ruins and the restoration of the remaining buildings has to be integrated with the plans for incorporating modern water reticulation, sewerage and fire services, underground electricity grid and stormwater systems within the historic settlement. The construction of an impressive caravan park well outside of the visible bounds of the historic site within an equally impressive natural setting has resolved one of the earlier use conflicts. However, even though the caravan park site was 1.5 km from Port Arthur it proved to have remains of the convict garden buildings. This entailed contracting for archaeological services to redirect and where necessary mitigate the effects of the development.[13]

All major issues have been considered in the Port Arthur Historic Site Draft Management Plan. That Plan for the most part details the activity of the Port Arthur Conservation and Development Project as well as considering National Parks and Wildlife Service management concerns. To some extent the Plan is a balancing act, establishing immediate priorities as well as setting the scene for future developments. The overall aim is to make the history of Port Arthur available and understandable as well as to ensure that the heritage qualities of the place are not eroded by natural or human forces. (*Summer 1982*)

Rippon Lea

Miles Lewis

Miles Lewis, Senior Lecturer in Architecture, Melbourne University.

Top Rippon Lea in 1880. The south west front as designed by Reed & Barnes in 1868, and slightly modified in 1880.

Above The drawing room created by Lloyd Tayler & Fitts in 1897, photographed c. 1902. (Photograph album held by the National Trust of Australia (Vic.))

The mansion 'Rippon Lea', in the Melbourne suburb of Elsternwick, came as something of an embarrassment to the National Trust in Victoria. It was not the house for which the Trust had battled so hard, but the grounds, which even on their present reduced scale are an unrivalled suburban estate of nearly six hectares (fourteen acres), with a large ornamental lake and rustic bridges, cascade, artificial mount and lookout tower; also a gigantic fernery, various lesser structures, and sweeping lawns about which peacocks now strut with a proprietorial air.

The house was a different story. It had been begun in 1868 as a substantial, but not grand villa in that patterned polychrome (or multi-coloured) brickwork which most people even today find a little too strident. It had been extended at various dates by the original owner, Sir Frederick Sargood, and after a few years interregnum had been bought by Benjamin Nathan, proprietor of the Maples furniture store chain. Nathan had made his own improvements, and after his death his daughter, Mrs. Louisa Jones, had made more, in what is really a rather splendid Hollywood style. Mrs. Jones it was who enlisted the Trust's support to fight the Commonwealth Government's plans to acquire large parts of the estate for the use of the Australian Broadcasting Commission, and Mrs. Jones it was who ultimately bequeathed the whole property to the Trust.

What can one do with a house that has altered so much over time, and should it be displayed to the public at all? These questions have begun to be answered by the scraps of information which have gradually come to light since the Trust took possession of the property in 1974 — evidence which has if anything enhanced the architectural importance of the original house, while also helping us to understand and appreciate the later extensions and alterations. We know that Frederick Sargood senior, who was in Victoria only from 1850 to 1858 while he established the drapery business which allowed him to retire to England, and which provided the foundation of his son's substantial fortune — that Sargood senior leased in his retirement 'Broad Green Lodge' at Croydon, with its own landscaped grounds and lake. It was surely the son's visit to this estate in 1861 which inspired the conception of the Rippon Lea gardens a few years later and also, it now seems, had something to do with the design of the house.

Overall, the character of the house must be attributed to the architect Joseph Reed, who had visited Italy in 1863 and returned to launch upon a startled Melbourne his polychrome brick style with round-arched openings, inspired by the Romanesque work of Lombardy. The firm of Reed and Barnes was well known to Sargood for in 1863 they had designed warehouses for his firm in Flinders Lane, Melbourne, and Dunedin, New Zealand, and one of their pioneering essays in the Lombardic Romanesque was Peter Davis's house in St. Kilda, quite close to where the Sargoods then lived. Thus we can understand why Rippon Lea came to be designed in this same extraordinary style, but it is less obvious why a

two-storeyed window bay, a feature quite alien to the style, was introduced as a prominent element of the south-west or garden front. The answer, we may now surmise, is that Sargood had seen such a bay at his father's 'Broad Green Lodge', a house built in the 1830s, when such an element would have been common.

The history of the nineteenth century extensions to the house has always been rather obscure, but it has been supposed that they were carried out by Reed and Barnes in about 1880–1882, when, after the death of his first wife, Sargood visited England and returned with her successor, and in 1887. We can now deduce from a combination of the evidence of old photographs, drainage plans, municipal rate books and family reminiscences, that the last work by Reed and Barnes was probably done before 1880, and that major alterations including the addition of a tower and a second storey to the back wing, as well as the enlargement of the lake, were carried out by other hands from 1882. The architect may well have been Lloyd Tayler, who was certainly doing some work on the property by 1889 and whose firm, Lloyd Tayler & Fitts, was responsible for major and reasonably well-documented work in 1897.

It was the work of 1897, hitherto unresearched, which can now be seen to have transformed the house into a mansion. The front hall was widened and completely reconstructed, the study lengthened into a drawing room measuring 40 by 24 feet (12 by 7 metres), the bathrooms enlarged, and a conservatory and porte-cochère added. It was not, however, the size, but the quality of this work that counted. The columns of the portico were of imported Peterhead granite, those in the entranceway

Left The drawing room as redecorated by Maples before 1918, a contemporary tinted photograph. (*Real Property Annual*, 1918, p. 46)

Right The south west front as it is today.

of Belgian St. Anne marble, and those in the hall from Spain, while slabs of lovely grey marble were used in the downstairs washroom. The reception rooms were hung with 'the best artistic decorative paper ... of the most modern designs' procured by the architects from Europe. Electric lighting was introduced, and the fittings designed by the architects were specially manufactured in England, and the same process may have been followed for the conservatory which, though apparently tailor-made for the property, bears the brand of George Smith's Sun Foundry of Glasgow on the castings.

At last, therefore, we have a reasonably complete picture of how the house reached its peak of grandeur in Sargood's time. We also know more about the twentieth century work. An illustration published in 1918 shows the drawing room, modernised once again from the form it reached under Lloyd Tayler & Fitts in 1897: it is not named, but labelled 'A modern drawing room — the furnishings and decorative plan suggested and carried out by Maples, Prahran'. In other words, Benjamin Nathan had had his own staff convert the room into a modern showpiece, and it was obviously a work of some significance in its own right. It was before the discovery of this illustration that the Trust had already re-redecorated the room in something approaching its Victorian form, so that Nathan's showpiece has been lost. One wonders whether, when this house comes to be fully researched, the Trust may not discover the twentieth century work to be quite as important as that which preceded it, and also, it may be added, a great deal easier to restore with confidence. (*Winter 1982*)

The restoration of Clarendon

G. M. W. Clemons

Dr. G. M. W. Clemons is the Chairman of the Clarendon Restoration Committee, and author of the section on Clarendon in Historic Homesteads of Australia, published by the Australian Council of National Trusts. He was the Tasmanian representative on the Council from 1962 to 1981.

William Cox with his wife and four younger sons arrived in Sydney on 11th January 1800 on the *Minerva* as paymaster/captain in charge of a detachment of his regiment the N.S.W. Corps. He had previously travelled to Sydney in 1797.

It is of interest that three of his sons were to build houses of great architectural merit, Hobartville (William), Clarendon (James), and Fernhill (Edward). There is no evidence that the family had any particular knowledge or interest in architecture and it has been suggested that they were fortunate to have an architect of outstanding merit. Despite research there is no evidence that an architect was employed for any of these houses although there is the very tenuous suggestion that Francis Greenway was involved with Hobartville.

James, second son of William, arrived in Sydney on the *Experiment* in 1804 at the age of 14 years. In 1814 he came to Van Diemen's Land and in 1819 was granted 400 acres in Morven where he was to build Clarendon which was completed in 1838.

In 1962 it was offered to the National Trust of Australia (Tasmania), to restore and manage. The house was obviously dilapidated, deserted (it had not been lived in since 1946) and with signs of structural damage. Consultations were held with the Premier Mr. E. Reece, and Mr. Roy Fagan. Satisfied that the Government would give some financial support, the Trust accepted the offer. It must be recorded that there were some who disagreed on the grounds that the restoration was beyond the capability and means of the Trust and in any event the house was not worth it. It is then fair to say that subsequently the Trust's view was amply vindicated by such acknowledged experts as Ian Lindsay (Edinburgh), Lord Euston (London), Professors Bourke (Melbourne) and Freeland (Sydney). In 1963 Professor Ernest Connolly of the U.S.A. came to Australia and

Drawing by W. Hardy Wilson. (Photo: National Library of Australia)

The restoration of Clarendon

was taken to Clarendon where he made a detailed examination. It was then arranged for him to see Mr. Fagan, to whom he subsequently wrote:

From a professional point of view, what impressed me most in Tasmania was the number and quality of your old buildings. Your early architecture is very distinguished and a justifiable source of pride to the State. In fact, of all colonial buildings in the Commonwealth known to me before my visit there was one in particular I most wanted to see and that was Clarendon. I knew it initially from the beautiful drawing by Hardy Wilson, including its handsome colonnade.... You can perhaps, imagine my disappointment when I saw the building with the colonnade missing, and the fabric deteriorating ... After examining the building rather thoroughly I was convinced that the importance and magnitude of the restoration transcends the scope of private undertaking ... As a friend of your preservation movement, I should like to see an Australian historic restoration carried out to the degree of thoroughness and accuracy which, through governmental sponsorship, we now accept as normal in the United States. Among the obvious benefits of such a restoration are the further stimulation of private preservation effort and the establishment of an ideal standard for such work.

These extracts leave no doubt of Professor Connolly's opinion and must surely have influenced the Government to give support.

The Trust then was committed to the restoration of Clarendon. I think the point should be made that when the Trust undertakes a work of preservation and restoration, work which is being done for the benefit of this and future generations, particularly when of the magnitude and importance of Clarendon, it must not be entrusted to the enthusiastic amateur or professional unversed in the methods, materials and skills of eighteenth century architects and builders.

Restoration has proceeded in stages depending on the availability of finance. Initially the problem was one of preservation. At some stage in the last century, perhaps *c.* 1880, it became clear that the house was sinking. The foundations were unable to support the great mass of the building surmounted by an extremely heavy parapet of large freestone blocks and the great weight of the portico. The solution adopted was to block in the windows of the semi-basement and fill it and the surrounding area with sand, at the same time removing the portico and parapet. This work was only partially successful as, when the house was accepted by the Trust, it was obvious that subsidence was still occurring, although gradually. The north-west and south-west were the areas mostly affected. It was plain that in order to preserve the house for the foreseeable future, complete underpinning of the foundations was an essential first step. This and the complete rebuilding of the collapsed terrace was accomplished in 1963–64. Repair of external walls and external painting, as well as work in the nine-acre park, was carried out in 1965 and partial restoration of the service wing in 1966.

The restoration of Clarendon

Above The dining room.
Right The south steps.

In the late 1960s a rather desperate position had been reached. While much had been done, most of the work was on the foundations and the exterior. Basically the house had been preserved but not restored. All funds received from the Government and those raised by Trust committees were exhausted. In such circumstances it is perhaps not entirely surprising that the suggestion should be made by some that the house should be disposed of.

It occurred to me that if some interior decoration could be done it would be possible to have fund raising activities in the house and it could also be let for other functions. With this object in view I approached six individuals, explained the position and with some trepidation asked for financial help. No-one refused and £875 was generously donated. With this money the main first floor rooms and the hall were decorated and fund raising activities held. It is my opinion that had this not happened Clarendon may well have been lost to the Trust.

The Trust's objective to restore the building completely and open it to the public was not forgotten. Furniture was gradually acquired through loans and gifts from Trust members and loans from the Queen Victoria Museum and Art Gallery, Launceston, loans from the Victorian Art Gallery and from bequests. Clarendon was first open to the public on Sundays in 1969, in 1972 for five days a week, and in 1975 for seven days a week.

The next major work was to restore the portico and parapet, but finance was still a problem. Since 1970 Mr. R. M. Green and I had known that there was a possibility of obtaining funds from the Australian Council of National Trusts. Then in 1973–74, $50,000 was given to the Australian Council of National Trusts by the Federal Government for the purpose of restoration and to be distributed by the Council at its discretion. All States were invited to submit applications. (Victoria and Western Australia did not make submissions.) The Council meeting at which the decision was made resulted in a protracted debate but eventually Tasmania was given $35,000.

A great step forward was made with the building of the portico and parapet in 1974–75. Ian Lindsay must have been delighted as, after seeing Clarendon, he wrote 'I am all for the great portico being restored as soon as possible, this very special feature . . .'

In 1975–79 with grants from the Commonwealth National Estate Program the work could go ahead on the park and establishing the original road and carriage circle, restoring staircases to the semi-basement, restoring the area walk, replacing windows and doors to all rooms in the semi-basement, flagging the passage, restoring the kitchen and finally completing internal and external painting.

This summarizes the work which has been paid for. During the years a

The hall, which extends from the north front to the south front with similar doors and fanlights at each end.

vast amount of work has been carried out in a voluntary capacity by members of the Trust.

There has been no further restoration since 1979, as no funds have been available. The restoration of the major portion of the semi-basement still remains to be done. The semi-basement housed the domestic offices, kitchen, pantry, servants' hall and the staff sitting and bedrooms.

The description of country houses by architectural historians and the connoisseurs of furniture usually neglects the domestic offices. Since the publication of Mark Girouard's *Life in the English Country House* (1978) and more recently *The Servants' Hall, A Domestic History of Erddig* by Merlin Waterson, this aspect has been recognised. Waterson points out that when something had to be demolished, the domestic offices were the first to go, together with their contents.

Of great interest to the social historian is the relationship between the family and their staff and an indication of this can be seen in the plan of the house. In some cases the domestic offices were entirely separated from the main house, in others, as is the case at Clarendon, they were part of it. Further, no special measures were taken to prevent views of the garden or the main entrance from the domestic area, as in the case of Clarendon. This points to the fact that although there was certainly segregation of family from staff, they were nevertheless one community. This important aspect of life at Clarendon cannot be recognised unless the semi-basement is restored.

The restoration of Clarendon has been a prodigious task and one may be permitted to say, a considerable achievement by the Tasmanian Trust. Should the Trust fail to complete this final and extremely important stage, history would judge it harshly. (*Winter 1983*)

Australia in Trust

Along the Ghan track

John Wood

John T. D. Wood, B.Arch., F.R.A.I.A., has recently undertaken research into tourism and conservation and the wider promotion of national estate concerns. He is a member of the Australian Branch of ICOMOS, and a life member of the National Trust.

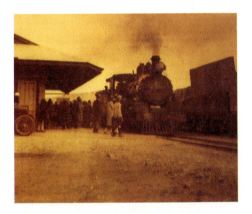

Above Alice Springs, August 1929. The first 'Ghan' train arrives in town. (Photo courtesy Mrs M. Byrnes)

Alice Springs, Ewaninga, Polhill, Ooraminna, Deep Well, Rodinga, Maryvale, Bundooma, Engoordina, Mt. Squire, Rumbalara, Musgrave, Finke, Crown Point, Duffield, Wall Creek, Abminga, Bloods Creek, Ilbunga, Pedirka, Mt. Sarah, Alberga, Wire Creek, Oodnadatta, Mt. Dutton, Algebuckina, Peake Creek, Warrina, Edwards Creek, Duff Creek, Box Creek, Anna Creek, William Creek, Irrapatana, Strangways, Beresford, Coward Springs, Curdimurka, Bopeechee, Alberrie Creek, Wangianna, Callanna, Marree.

As you read these names you start to pick up a rhythm not unlike that of the old 'Ghan' that, ever so circumspectly, rattled its way through these stations on the northern section of the Central Australia Railway until its closure in 1980. Yet, even more strongly, the names seem to capture the essence of this grand folly in railway construction with their mixture of Aboriginal, European and invented origins. For the railway that was pushed through the arid regions to the north-west of Herrgott (Marree) opened up country that only Aborigines and a few European explorers, graziers, missionaries and surveyors knew.

Commenced on 19 January 1878, the first stretch of the line from Port Augusta through the Flinders Ranges to Government Gums (later named Farina) was very much a response to the pressure of the South Australian Government from the miners of the Flinders region and the farmers. These farmers had taken up all the land south of Surveyor-General Goyder's line of safe rainfall in the 1850s and 1860s, and wanted to cultivate their wheat lands right to the Northern Territory border. Spurred on by unusually high rainfall, and despite Goyder's dire warnings, the Government succumbed and allowed land to be taken up on credit-selection to the border. The wheat farms were to be short-lived and with the commencement of drought conditions in 1879, crops failed and the surface soil literally blew away. The momentum of the railway was nevertheless continued and with the new demands for extension to the line coming from graziers in the north and in south-west Queensland, the line was extended to Marree in 1883.

The direction the line should now take became a matter of considerable dissension. As most of the cattle coming to Adelaide originated in Queensland there was substantial lobbying to take the line to the Northern Territory via Birdsville and Camooweal. Ultimately it was probably colonial rivalry which determined that the railway would unite the Territory with the main colony within South Australian borders.

The railway, now known as the Great Northern Railway, reached the Peake in 1889 and was opened to Angle Pole (Oodnadatta) on 7 January 1891 where it stayed until 1927. In 1907 the South Australian Government entered an agreement with the Federal Government for the surrender of the Northern Territory and the sale to the Commonwealth of the Great Northern Railway. It was not until 1926 that the Commonwealth Railways finally took over management of the railway and construction of the line to Alice Springs commenced on 21 January 1927, finally being completed in August 1929.

The feelings acquired from travelling in the Ghan were entirely a matter of individual perspective. If one was using the railway simply as the means for getting to Alice Springs it must have been a frustrating experience. If, on the other hand, one's motive was to undertake a unique railway journey and to discover some new horizons, it could be the most romantic of discoveries.

To travel by road, down from Alice Springs to Marree, along the now closed line, is quite another experience. For me it was a childlike journey through a vibrant land populated by ghosts, an almost continuous stream of new feelings, awe inspiring sights and extraordinary personal revelations.

Though a map of Australia would lead you to think you are passing through the middle of almost nowhere, the landscapes that unfold on the journey represent not just the broad expanses of scrubby plain, nor the rolling series of red sandhills, but also the most intimate of scenes. Such are the places where the settlers established their confined and isolated worlds.

The first of these is at Deep Well some 78 km south of Alice Springs. Focusing on the well built by William Hayes around 1900 this 'settlement' consists only of a standard Commonwealth Railways' house constructed about 1930 and the ruins of a stone rubble-walled house built by a person named Johansen *c.* 1915 and the fine stone head of the 66-metre-deep well. Nothing startling, but like all these now deserted outposts, the sum of the efforts of those who worked and lived here has left a distinct 'presence', along with the more tangible detritus of settlement scattered over the stony fields. The timber Railway house with its large corrugated iron roof, screened verandahs and fence built from sleepers driven vertically into the ground, was still intact, the only such example between the Alice and Oodnadatta, a fact which is of some importance for conservation action given that this line had only been closed nine months before.

Thirty-six kilometres further down the track we came to 'Maryvale' homestead, a shady collection of buildings spread beside the Hugh River which rises in the James Ranges and joins the Finke to the west of Engoordina. This marked our only real deviation from the line of the Ghan as we drove to the south of west to that most remarkable central Australian signpost, Idracowra as it was known by Aborigines or, as John McDouall Stuart named it on 6th April 1860, Chambers Pillar.

Approached across a series of 6–11-metre-high red sandhills studded with acacias and thryptomene, it is a stunning event to suddenly come across this column of sandstone, sitting atop the hill which has been eroded from it, in the midst of a spinifex and mulga covered plain. When he first sighted it Stuart described it as 'a remarkable hill, which at this distance (12 miles) has the appearance of a locomotive with its funnel'. Ernest Giles later indulged in a little more graphic prose:

'By this time we were close to the pillar: its outline was most imposing. On reaching it, I found it to be a columnar structure, standing upon a pedestal, which is perhaps eighty feet high, and composed of loose white sandstone, having vast numbers of large blocks lying about in all directions. From the centre of the pedestal rises the pillar, composed also of the same kind of rock; at its top, and for twenty or thirty feet from its summit, the colour of the stone is red. The column itself must be seventy or eighty feet above the pedestal. It is split at the top into two points. There it stands, a vast monument of the geological periods that must have elapsed since the mountain ridge, of which it was formerly a part, was washed by the action of the old Ocean's waves into mere sandhills at its feet. The stone is so friable that names can be cut into it almost to any depth with a pocket-knife; so loose, indeed, is it, that one almost feels alarmed lest it should fall while he is scratching at its base.'

Chambers Pillar thus became both a pointer for all those who subsequently travelled to the north before the advent of the railway and a monument which recorded their names. There, cut deeply in the stone, in a variety of styles, are John Ross who surveyed the route for the Overland Telegraph in 1871, Basedow the geologist, Frank Wallis who ran one of the earliest stores in Alice Springs, sundry Hayes family members who settled in the region and the infamous W. H. Willshire. Mounted Constable Willshire's efficiency in wiping out Aborigines in any area of Central Australia and the Northern Territory from which came reports of cattle killing, was matched only by his liking for etching his initials in rocks and trees from Port Augusta to the Victoria River.

It was, however, another name, which most moved me, that of 'B. C. Besley, 1890'. Besley is my mother's family name, and there was a completely unexpected visible record of my very own link with central Australia through my great-great uncle, Brian Charles Besley who had been Inspector of Police for the north of South Australia. Without wishing to labour the point, a journey which was already a marvellous education, had added to it another dimension.

On the return to the track down the line we detoured via the ruins of Mt. Burrell homestead. Mount Burrell was originally held by a lease granted in 1875 and was one of the first pastoral properties to be stocked in central Australia. It soon changed hands and by 1884, when the homestead was built, it was in the hands of Sir Thomas Elder who used it predominately for breeding horses for the Indian army. It was later taken over by the ubiquitous Hayes family before 'Maryvale' was established. All that remains today are two fine fireplaces and chimneys and a wonderful litter of china and pottery fragments and metalwork from early implements on the site of the old smithy.

If you decide to follow the track down the Ghan from Alice Springs, there is one weekend in June you must avoid, when the track is closed for two days for the Finke Desert Race for motorcycles which churn down from Alice Springs to Finke on the Saturday and return the 225 kilometres on Sunday. The point of this aside is simply that as we were camped by the track near 'Maryvale', a bobbing, weaving light suddenly appeared and, drawn by our campfire, pulled up. The balaclavaed, goggled and leather suited apparition that approached us turned out to be a two-time winner of the race. Just as this country will suddenly provide you with an unexpected vista, so it seems will it produce extraordinary characters. Bob, the motorcyclist, had fled boredom in Melbourne at age 19 by hopping on his bike and riding until he came to the Territory. Now, eight or so years later he was assisting with management on 'Mt. Ebenezer' station on the road from the Stuart Highway to Ayers Rock. When he came on us he was reconnoitring the track that he was to race down the next day. Having blown his headlamp at the first attempt to leave our camp he stayed the night and carried on the traditions of oral history in the Centre with his own, and acquired stories of the region.

Alice Well, thirty-two kilometres south of Maryvale, is also situated on the bank of the beautifully wooded Hugh River. A major watering place on the main stock route to the south, it had been another Hayes family station and the site of the regional police station from 1911 to about 1928. All that now remains is the old stone well-head, similar to that at Deep Well, the one wall of a plastered stone-rubble building and the Government bore and tank which replaced the well and supplied water via a 2.5-kilometre pipeline to the railway at Bundooma. In addition, there are the remains of a pump-jack and the posts and pulley system that was camel operated to draw water from the well, a unique piece of Heath-Robinsonian ingenuity. Constable Jack Mackie was the policeman stationed at Alice Well, whose duty it was to travel the forty-three kilometres south-east to Horseshoe Bend in October 1922 to prepare the

police report on the death of the Rev. Carl Strehlow after his epic 260-kilometre journey down the Finke from his life's work at Hermannsburg Mission.

The next stretch is straight down beside the railway line to Finke. This cuts through the western edge of the Simpson Desert, and the generally flat sandy terrain was relieved by the undulating sandhills and a marvellous array of wildflowers. The country was far from the barren desert of our preconceptions, for a succession of good rainfalls over recent years had ensured rich growth throughout our trip. Predominant were native fuchsias including the Elegant Wattle (*Acacia victoriae*), colony wattle (*A. murrayana*), Gidgee or Gidyea (*A. cambagei*) and mulga (*A. aneura*); the showy grevilleas, (*Grevillea juncifolia* and *G. stenobotrya*); the eucalypts including Bloodwood (*Eucalyptus terminialis*) and Desert bloodwood (*E. setosa*); the lovely Corkwood (*Hakea suberea*); the pink thryptomene of the dunes (*T. maisonneuvei*); the beautiful hibiscus-like shrubs (*Gossypium australe* and *Radyera farragei*); the majestic Desert Oak (*Casuarina decaisneana*); and the cheerful variety of perennials including the purple wild tomato and the various yellow daisies.

Finke is a tiny township on the south bank of the river from which its name derives. Its significance was only established with the arrival of the railway in 1928, prior to which Old Crown Point to the north-west and Charlotte Waters to the south-east on the route of the Telegraph were the major foci for the district. Now that the railway is gone it is a sad little place notable only for its pub (which was closed) and a nice example of the Commonwealth Government's standard arid zone house built in the 1930s and serving as the Police Station. With its ripple iron sheeting, awnings which provide shading, breeze control and a reasonable barrier to dust storms, these buildings are an admirable example of adaptation for a testing climate.

Just to the north of the town are a series of pump houses at the site of the Government bore which was used for watering the trains, and nearby is the extraordinary railway crossing of the mighty Finke. It seems strange to describe a broad tree-lined stretch of sand thus, but those dry river beds of central Australia do impart a sense of their might. The Finke or Lira Beinta (Larapinta) as it was known to the Aranda, is the greatest of all. Drawing its sources from Mt. Sonder and Mt. Giles in the Chewings Range, this most ancient of the world's rivers cuts its way through the Macdonnell Ranges, crosses Missionary Plain then, by Mt. Hermannsburg, enters its long winding gorge through the James Range, meanders off to the south-east past 'Henbury', picking up the Palmer and Hugh Rivers, and finally strays out past the most easterly of properties to wash out in the Simpson Desert. No one who sees the Finke dry or in flood ever forgets it. Railway builders and travellers on the Ghan were no exception. The line was originally taken over the River on a high level bridge, in 1928, which was washed away by a flood two years later. The builders learnt their lesson and from then on, first on a stone bed and later on a concrete causeway, the line was taken across at the level of the river. When it flooded, the train could not cross, the line might disappear, and that was that! They simply cleared it when the waters subsided and relaid the track if necessary.

Leaving Finke to the only evident life in the town, an Aboriginal rock band practising in a galvanized iron shed, we travelled along the Stock Route through 'New Crown' station and south to the site of the renowned Telegraph station at Charlotte Waters.

The site was discovered on 10 January 1871 by two surveyors, G. R. McMinn and R. R. Knuckey, who were looking for a watered route for station sheep to supply the teams constructing one of the cental sections of the Overland Telegraph from the Stevenson to the Hugh. Knuckey, no doubt struck by the beauty of the waterbed named it after Lady Charlotte Bacon, 'Ianthe' to whom Byron dedicated 'Childe Harold's Pilgrimmage'. Knuckey also determined it should be the site of one of the four central telegraph stations (Alice Springs, Barrow's Creek and Tennant's Creek were the others). By December 1871 construction had commenced on the station buildings using a lightweight white sandstone from the Hearne Range 12 kilometres to the south. The repeater station buildings were quite impressive, built round three sides of a courtyard and containing the telegraph station, post office, store and Station Master's residence. In addition there were other buildings to house operators, linesmen, labourers and police.

The existence of the water hole may have determined that Charlotte Waters was an ideal situation for a telegraph station but it is difficult to imagine a more desolate spot for human habitation. As at Chambers Pillar, however, the spot had some powerful personal meaning for me. For here a great uncle, Francis Gillen, had been Station Master from 1876–1887 and my grandfather's half-brother P. M. 'Paddy' Byrne succeeded him for at least 12 years.

Gillen and W. B. Spencer who visited Charlotte Waters described the setting thus:

'The station is placed close to the northern edge of a wide plain. The main buildings form three sides of a quadrangle, the fourth side being closed in by strong gates — or rather it used to be in the early days when it was first built and it was necessary to have protection against the blacks. At that time the doors all opened on to the quadrangle, and every room had loopholes through which, if necessity arose, the officials could defend themselves. In the early days, when the northern railway only reached as far as Port Augusta, there were eight hundred miles of, for the most part, dry and often sterile country to be crossed between the head of the line and Charlotte

Waters. The distance is now reduced to 150 miles, and the time may come when the railway will be extended still further north; but as yet all these stations along the telegraph line are completely isolated from the outside world, and Charlotte Waters looks out upon a great open, stony plain without a sign of human habitation. North and south runs the line of telegraph poles, streaking away to the horizon, and the ticking of the instrument, as the messages pass through, only serves to heighten the feeling of isolation.'

No wonder that here Gillen commenced his study of Aborigines that was to lead to his important anthropological studies and to his appointment in 1891, while stationed at Alice Springs as Inspecting Officer of the Central Section of the telegraph, as the Sub-Protector of Aborigines.

Today the isolation is complete. With only the rubble of the station buildings, a tank, some grave railings and a few posts protruding above red gibber plain, all is still. Only the line of the Hearne Range on the horizon gives a token that the world extends beyond this place.

Telegraph operations transferred to Finke in 1930 and the Post Office officially closed later that year. The police station and telephone services continued until 21 August 1938 when the station was officially closed. After that the condition of the buildings, already poor, deteriorated. The fate of an historic post was settled when, after tenders for its purchase were called around 1945, the roofing iron and timbers were removed for use at 'Andado' homestead to the north. In 1947 the walls were demolished and the stone used for the construction of a house at 'New Crown' thirty kilometres north.

That travel can still have its chancey side was nowhere better illustrated than on our next section. Continuing down to the railway at Abminga we intended to do a loop to the south-east to take us via the ruins of 'Dalhousie' homestead and thence down to the Stevenson and back to the railway at Pedirka. Even with good maps the tracks around Dalhousie are mostly marked 'position doubtful'. Indeed they are. However, we thought we were in luck when we came across the extraordinary sight of a collection of tents, vehicles and about a dozen light aircraft! The occasion was a 'fly-in' of people from New South Wales, South Australia and the Northern Territory for a weekend of fun. Contrary to popular tradition, the travellers were not welcome when we sought clarification of the route we wished to follow. On being informed that we were heading for Marree, our 'guide' said to keep on our current track. Sixteen kilometres of heading due east quickly raised our suspicions. We were indeed heading for Marree, but via Birdsville and the Birdsville Track. The 'right track' we were on was taking us straight across the Simpson Desert to Poeppel's Corner, the junction of the Queensland, Northern Territory and South Australian borders! Further enquiries from others at the 'fly-in' were no more helpful, so we retraced our path to Abminga and then down across The Hamilton and The Alberga to Oodnadatta.

Largely, I suppose, because my grandfather who had also worked with the telegraph had often referred to it, and because it is such a wonderful name, I always expected Oodnadatta to be a sizeable town. I suppose that, in comparison to Finke, it is. Certainly there are some fine buildings, not least of which is the stone railway station building constructed shortly after the railway arrived in 1891. Also of interest, and built about twenty years later, is the barrel-shape roofed, corrugated iron goods shed with its internal tensile ties which enabled the structure to flex in high winds. As the railhead stayed at Oodnadatta for thirty-six years, it developed a considerable importance as the main dispersion depot for central Australia, with about 400 camels running the strings that followed the Overland Telegraph as far as Newcastle Waters, 1160 kilometres to the north.

About 30 kilometres south of Oodnadatta, the road rejoins the line of the Ghan and stays with it for the rest of the distance to Marree. There

Along the Ghan track

Chambers Pillar, during Spencer & Gillen's trip, 1895.
Shifting camp, 1928.
Strangway, Springs Post, 1886.
Loading camel wagons, 1928.
The rail head, North South line Adelaide to Darwin, 1928.
Charlotte Waters Telegraph Repeater Station, c.1922.
Hospital, 21 mile camp, 1928.
Staff tents, 21 mile camp, 1928.
Abminga Bore, 1928.
The Ghan passenger train approaches Alice Springs.

is little to record in the next 230 kilometres save the most substantial of the bridges on the railway, that at Algebuckina, a twenty-three span bridge across The Neales which was built in 1889. At this stage the line is running parallel to and about 100 kilometres to the west of Lake Eyre North. The line at last starts swinging around to run south of the great lake at Box Creek. The only habitation between Oodnadatta and Marree on this road is the William Creek Hotel, built in 1889 and still, thankfully, operating.

About forty kilometres on, situated on a rocky hill above the road, is Strangways Springs. Discovered by P. E. Warburton in 1858, the mineral springs that bubbled out at this site formed the base for a pastoral block which was first started in 1863. A station house was built in 1867 from the local travertine together with a store, a smithy, yards and woolshed. By late 1871 the Overland Telegraph had reached Strangways and part of the pastoral lease and the buildings were resumed in 1872 for the repeater station and telegraph office which opened in August of that year. Because of the rocky foundations on which the buildings were located, underground tanks could not be provided and, instead, a substantial roofed stone above-ground tank was constructed by 1884. With the railway approaching, a police station was added in 1886. However the railway station and various other buildings, including the inevitable hotel, were located by the line some two and a half kilometres from the telegraph station. Nothing now remains of the railway 'town'.

The telegraph line was moved alongside the railway in 1887 and the telegraph office was closed in 1896. There are some substantial remnants of the buildings including the repeater station, the kitchen and men's living quarters and the woolshed. The feature, however, is the tank which, though no longer roofed, is substantially intact. It is unlikely that it will be for long.

Our heritage embodied in these remote sites is constantly at risk. Much of it has been systematically demolished by those seeking building materials that are hard to come by in such isolated areas. Charlotte Waters has virtually disappeared, Strangways is rapidly following and so are the railway workers' quarters dotted along the line and only recently abandoned. The roofing iron had gone from all those we saw with the exception of Curdimurka, 64 kilometres further on from Strangways. These stone-walled skillion roofed structures, whilst no more than functional in appearance, are, nevertheless, important examples of the railway vernacular. Another unusual relic at Curdimurka is a thirteen-metre-high desalination plant, used to clean the bore water for the engines, lurching alarmingly towards its storage tank.

From Curdimurka the line passes close by the shore of Lake Eyre South and it was here that the railway was threatened by the great flooding of Lake Eyre in 1974. The embankment which carried the track was actually raised one metre. The water covered the adjacent road in September and wave action started to erode the embankment. Slabs of rock were brought in by 300 train loads from Finniss Springs and dumped on the lake side of the embankment, which fortunately held, although it is reported that the Ghan got a good washing down from the breaking waves!

Herrgott Springs, two and a half kilometres north of Marree, were discovered by J. A. F. D. Herrgott, a member of Stuart's party, and named by the latter on 15 April 1859. Stuart described these mound springs thus:

Seen from a little distance these springs, at which we camped, resemble a salt lagoon covered with salt, which however, is not the case; it is the white quartz which gives them that appearance. There are seven small hillocks from which flow the springs; their height above the plain is about eight feet, and they are surrounded with a cake of saltpetre, but the water is very good indeed, and there is an unlimited supply.

As Ian Mudie points out in 'The Heroic Journey of John McDouall Stuart' it was some time before it was realised that springs such as Herrgott and Strangways marked a fault line that allowed water up from the Great Artesian Basin bringing with it minerals that were deposited to form the mounds.

It was around the springs that the original settlement developed, for it marked a continuous chain of permanent water which allowed optimistic squatters to move north of the Flinders Ranges. As properties were settled beyond Herrgott Springs, so it became a natural centre, and stop, for stock being sent south to markets. By 1880 it had become a major depot for the cameleers who had entered the transport business with Sir Thomas Elder and his partner Stuckey. These strings operated up and down the Birdsville Track (then known as the Queensland Road), the Overland Telegraph line to the north, the Strzelecki Track to Innamincka and east through Blanchewater (Lake Blanche) to New South Wales.

With the arrival of the railhead the town moved across from the springs and the bore, and a substantial settlement developed. The two-storey coursed random rubble hotel with its beautiful balcony running the length of its eastern facade is the most outstanding example of the flurry of construction around 1884. But the stone cottages and even the later corrugated iron houses have pleasant proportions and, as far as can be determined, were functional enough to help the residents survive the staggering summer heat that Marree endures. One of Stuart's companions on an 1860 expedition recorded 128°F in the shade and 173°F in the sun!

The importance of Herrgott as a stock trucking centre, once the railway was running, is exemplified by the growth in carriage of cattle and sheep. In 1884, 1,946 cattle and 1,180 sheep were carried south, by 1892 this had risen to 15,651 and 20,093 respectively and by the 1920s, although sheep had virtually disappeared the line was carting nearly 40,000 cattle.

No better illustration of the *raison d'être* for Marree can be given than the sight of the railway line running through the middle of the town supplanting the traditional dusty main street. Marree was spawned by the railway's coming, it is dying with its departure.

It was here that we left the line of the Ghan and turned left up the Birdsville Track. Just as the Overland Telegraph line, over a hundred years ago, was a constant guide, companion and reassurance to the pioneers of Central Australia so was the Ghan line to us. It may seem strange to attribute such qualities to inanimate steel, hardwood sleepers and stone ballast, but this line has affected many lives and in so doing it affects us and brings us a little closer to an understanding of this country. An understanding that Aborigines have possessed for thousands of years.

In exploring these remote places of Australia, we still cling to the slender threads of communication that our forebears established. But it can give us the self-confidence to venture into those vast areas that so few white occupants of this land have ever seen or really believe exist except on maps or in coffee-table books. It is only then that the true magnitude and wealth of the heritage which is entrusted to us will have its real impact. (*Winter 1982*)

Australia in Trust

The art of the carpenter

Frank Bolt

Frank Bolt is a well-known Tasmanian heritage photographer. He has published several books on local heritage subjects, including his recent 'Old Hobart Town Today'.

Much of our heritage literature concerns itself with fine buildings and famous builders, yet the remaining evidence of the past is full of examples of fine workmanship on the part of the common tradesman, the (usually) semi-literate worker who travelled from job to job with his chest of tools, and who needed only the merest outline of what was required to get him started. One such a tradesman was the carpenter, and especially in Tasmania with its abundance of good-quality timbers the contribution of his trade to the general appearance and utility of the structures he built is quite notable.

During the first few decades of settlement the supply of tradesmen was largely a matter of what could be found among the convicts, but later on was supplemented by the arrival of free tradesmen, many of whom were extremely well trained and qualified. Nevertheless, they all were confronted with the unfamiliarity of a new country: new kinds of timbers, the lack of nearby and well-stocked hardware stores, and the much greater range of climatic conditions to which the final workmanship had to be exposed.

Yet they very rapidly discovered the beauty of cedar, the strength of hardwood, the artistic possibilities of Huon pine, the utility of imported planking, the right profile for weatherboards, and the use of lath as a base for plaster. And with these elements well under control and with a great degree of freedom with regards to style, these carpenters soon developed their own and often flamboyant ways of dealing with structures, shapes and detail.

The earliest remaining examples of carpentry in Tasmania were mainly

Left The bush church at Kunnarra, dating from 1874.

Right Some of the older townships in Tasmania still contain quite a few buildings which can easily be recognised as early shops, like this one near Huonville.

Australia in Trust

Centre Although designed by the well-known Hobart architect, Henry Hunter, the detailing of this gate in Battery Point, Hobart, nevertheless follows quite traditional lines.

Right Designed by John Lee Archer and built by convict labour in 1834, the construction of this magnificent roof high up in St Johns Church, Newtown, would have been a major feat of engineering.

Middle row

Far right One of the many doorlights that can still be found in the Hobart of today. They all date from c. 1840 to about 1855, and came in an endless variety.

Centre This shopfront in Colebrook has not changed much during the past century. In order to keep the cost down, the carpentry of the windows and doors was kept to a minimum.

Right A beautifully proportioned window with its obligatory holland blind.

Bottom row

Far right A window airing a barn on the Tasmanian east coast. A simple but very cleverly thought out arrangement of a stick with notches moves these levers up and down as required.

Centre An early shop window in Oatlands, Tasmanian midlands.

Right Kuranui, a superbly designed, built and preserved timber cottage overlooking the mouth of the Tamar River.

Below A building near Orford on the east coast of Tasmania, which would be instantly recognised by any rural worker as a farm building housing a wool press. Both the structure and its detail were kept extremely simple.

Top row

Far right Much altered by 'repairs' in later years, this roof window in Hobart is of interest because of its wildly elaborate decorations. The style and detailing is typical of Hobart c. 1880–90.

The art of the carpenter

Australia in Trust

produced by convict labour: very structural, and with no nonsense such as artistic details, thank you! A greater sophistication became apparent during the late 'thirties and 'forties when skilled craftsmen began to arrive, and from the middle of the century onward quite a considerable number of them were available all over the island, skilfully using the local timbers as a familiar material to fashion those structures which the settlers asked for: homes, churches, storehouses, shops, mills and bridges.

Few things were impossible to these people who moved their tools with a will; their contribution towards the heritage of this island is significant, and it was not until the 1890s that their influence began to wane when a new generation of architects and engineers took over the design functions within the building industry.

Of course, timber is not everlasting; it burns and breaks, and is easily tampered with, and as a result the number of fine examples of the art of the Tasmanian carpenter is shrinking all the time.

The photographs reproduced here show some of the items fashioned from timber which caught my eye over the past few years, and hopefully will encourage others to take a greater interest in this fascinating part of our heritage. (*Winter 1982*)

The art of the carpenter

Below The timber verandah of Willow Court in New Norfolk, a hospital for the mentally ill, dating from the early 1830s and designed by John Lee Archer.

Top left One of the most remarkable places of worship in Tasmania was this Presbyterian 'church' in Tunnack, built by an ex-convict, tactfully known locally only as 'Joe the Splitter'.

Bottom left A finely decorated shop in East Devonport, possibly dating from 100 years ago.

Adelaide's stone buildings

Alan Spry

Alan H. Spry is a senior consultant at the Australian Mineral Development Laboratories, South Australia.

The Adelaide Gaol. (Photo: National Trust of S.A.)

More than in any other city in Australia, the varying use of stone in the buildings of Adelaide reflects the social and economic aspects of this city's history over a century and a half since its establishment in 1836. The influences of boom and depression, of affluence and economic stringency, of the availability of Government grants for religious or educational purposes, of transportation and of fashion are seen. The large proportion of stone buildings partly reflects the abundance of that material and the lack of suitable timber, but mainly results from a legislative peculiarity as the Building Act of 1858 banned timber buildings because of fire risk. Strangely, all of South Australia's building codes up until 1940 were basically derived from the English 1667 Act aimed at regulating the rebuilding of London after the Great Fire.

There is no shortage of brick buildings in Adelaide and its suburbs, as brickworks were operating in and around the city by 1838, within two years of the original settlement, but nevertheless it is stone, in humble domestic dwellings and the grander public edifices, in city and in suburbs, which gives Adelaide its distinctive character.

It is possible to recognise a number of distinct periods which were dominated by certain types of stone and an historic thread may be followed from the earliest practice when surface and near-surface stone was derived from dozens of sources (including the building site itself), with a peak of activity with stone from a few large quarries treated with superlative craftsmanship in the High Victorian boom-times of the 1880s, thence to the decline caused by depression late in the last century, and finally to the dwindling use of this expensive and demanding material in modern times.

Three main types of stone dominate Adelaide's older buildings: bluestone and limestone (which were not used to any degree elsewhere in Australia) and sandstone. Marble, granite and slate were also utilized but to a minor degree.

The bluestone of Adelaide (as distinct from that from Victoria and New South Wales which is basalt) is a dark-coloured (brown, grey, black and orange but never blue) argillite related to siltstone, slate or quartzite and is visible everywhere in nineteenth century buildings of all kinds. This type of stone was quarried in the 1830s and '40s at Green Hill, from the 1840s to late in the century at Brownhill Creek and Mitcham, from 1850 at Tarlee, and from the 1850s to the 1900s at Dry Creek and Auburn. The two major types were from Glen Osmond (1840s to 1920s) and Tapleys Hill (1870 to 1920s). The stone fragments are rather small and irregular in the older buildings whose random rubble walls contain a great deal of mortar, but in progressively later examples the blocks can be seen to be larger and better-dressed culminating in walling slabs up to 2 m long. A good example of the latter is in the 1925 wall of the College of Advanced Education in Kintore Avenue.

The hard white, grey, yellow and brown sandstones of Adelaide differ

Left
1 Bluestone is characterised by its dark colour and blocky nature. Construction in the last part of the 19th century utilised large, well-shaped blocks with sparse mortar whereas the older, random rubble contained blocks of irregular size and shape with abundant mortar.

2 The hard sandstones from the Mount Lofty Ranges were generally roughly dressed in domestic dwellings.

3 The tough Tea Tree Gully and Glen Ewin sandstones were finely dressed in some prominent buildings, such as the Supreme Court (1869).

4 Despite its cavernous appearance, the uncommon Adelaide limestone, here in the 1845 Trinity Church, is very durable.

5 Stone is still used in domestic dwellings in the suburbs of Adelaide. Construction involving a mixture of sawn, soft, coloured sandstones (Manoora, Basket Range, Macclesfield, Kapunda and Brinkworth) with prominent tuck pointing, is unique to Adelaide.

Right Edmund Wright House (formerly the Bank of S.A., 1875–78) marks the culmination of the use of hard sandstone of the Tea Tree Gully type.

somewhat from their younger counterparts in Sydney, Brisbane, Hobart and Perth in that they are weathered Precambrian quartzites which have been hardened, folded and metamorphosed, as distinct from the Mesozoic sandstones of the other States. Some are so tough and hard to dress that they were often only crudely shaped into blocks and left with an outer rough face, whereas the Sydney and Tasmanian sandstones could be finely dressed to geometrically perfect ashlar with the tightest of joints and smoothed or finely dressed outer faces. The most important South Australian sandstones were from the Tea Tree Gully area; their great durability is demonstrated in the preservation of the details of carving and the dressing marks made by pick, chisel or patent hammer after more than a century on the Supreme Court. Unfortunately, the Tea Tree Gully quarries ceased production in the 1920s and are surrounded now by the urban fringe so that extraction of stone, even for restoration work, is no longer possible. Softer yellow to light brown sandstones from various parts of the Mount Lofty Ranges were used in many residences between 1870 and 1960. Very soft multicoloured and figured sandstones are still used.

Limestone (brown, grey and white) is soft and easily quarried and dressed, hence it is cheaper than sandstone which it has come to replace. It has been widely used in Perth and Fremantle where it is the locally abundant stone and to a small degree in Melbourne, but only imported material is seen in Sydney, Brisbane and Hobart. Fieldstone (kunkar, travertine or calcrete), the unique Adelaide Limestone, and later, Murray Bridge, Waikerie and the very soft Mount Gambier Limestones (and pink dolomite) were all used in Adelaide.

Just as the social and the architectural history of the city can be divided into periods (Primitive, Colonial, Early, Mid, High, and Late Victorian, Federation, Modern and others), so can the use of the various kinds of stone. This is not surprising because utilization of any building material depends on labour, transport and technology, hence on the level of population, prosperity, optimism and development.

The earliest recognisable (Pioneering) period extended between about 1836 and 1840. A year after settlement began, Adelaide was a village of tents, timber, wattle and daub, transported and temporary structures, but by 1838 the first stone houses appeared. Limestone was used first so that walls were constructed of stone from the site and mortared with lime burnt on the spot. This was closely followed by bluestone, quartzite and slate from a large number of small scattered local sources. Round boulders of white nodular surface limestone (calcrete) occur across the South Australian plains and are seen in random rubble walls of buildings from Goolwa to Port Augusta and beyond to the Riverland, South-East and Eyre Peninsula. Calcrete was used until about the turn of the century in such buildings as Christ Church in North Adelaide (1848), the Almonds at Walkerville and a gallery building in Barton Terrace (1856). The scramble for this stone became so fierce that a regulation was passed forbidding its taking from parklands and reserves. However, it is rumoured that at least part of the stone for Christ Church was taken in the dead of night from Palmer Gardens opposite.

The other distinctive limestone whose use began in this early period is the Adelaide Limestone found now in only eight structures, all of great historic interest and including Government House (1840 and possibly 1838), the western wall of the Adelaide Gaol (1840), Trinity Church (1845), the Mounted Police Barracks (1853–8) plus adjoining structures and the old Legislative Council building (1855). The stone was excavated from the area between the parade ground behind Government House to the Adelaide Railway Station. The shortness of the period (?1838–58) of its use was probably due to the nuisance caused by its proximity to the city as shown by letters in 1847, 1848, 1849 and 1854 to the Colonial Secretary complaining of damage due to stones from the quarry falling on neighbouring properties. The quarry belonged to the Government and although it probably commenced operating before 1840, it was preceded by the privately operated bluestone quarry at Green Hill (near Beaumont) which opened in 1838 (its stone was used for 'Gleeville' and 'Tusmore Gate') and by the quartzite quarry at Stonyfell in about the same year. The Willunga Slate quarries commenced operation in 1840 and produced roofing and paving material for the next 80 years; much was exported (for example for roofing the G.P.O. in Melbourne and St. George's Cathedral in Perth).

There were 1,960 houses, shops and stores in Adelaide by 1841 including 500 stone or brick dwellings for a population of 8,000. A lull was caused by the depression of the early 1840s then there was an acceleration in building between 1844 and 1848.

Building between 1840 and 1850 was characterized mainly by the use of bluestone from the hills-face between Green Hill, Glen Osmond and Mitcham (seen in the Glen Osmond Toll House in 1841, 'Claremont' and 'Cummins' in 1842, 'Kurralta' in 1843, 'Benacre' and the Mountain Hut in 1845, Paxton's Cottage and Smith's Stores in 1846 and others). Thus the architectural Colonial Period corresponds to the first part of the Bluestone Period (1840–60).

The construction of the Adelaide Gaol included the use in 1843 of Tasmanian sandstone for copings, sills and towers. There may be more of this imported stone in Adelaide than is presently suspected.

One of the few major buildings of this period is the 1847 Magistrates Court. The source of the stone is not known and the standard of craftsmanship was extraordinarily high for the time, in fact, such smooth, tight, regular ashlar was not seen again in Adelaide for many years.

By 1850, South Australia's first mining boom (1840–51) was coming to an end, the population of Adelaide exceeded 10,000 and there were more than 4,000 brick and stone houses but there was an oversupply of labour and much unemployment in the building trade. The Mintaro Slate Quarry commenced operations in 1856 and is still working after 130 years, making it the oldest operating quarry in Australia. The opening of the Glen

Osmond Slate Quarry in 1850 marked the beginning of another decade dominated by the use of bluestone, so that the second or major part of the Bluestone Period corresponds to Early Victorian times. Large houses ('Beaumont House', 'Rath Gael', 'Tower House', 'Sunnyside House', 'Woodfield' and others) of this material appeared in the early 1850s. The middle of this decade saw a burst of building of bluestone churches (Scot's, 1851–6; the first part of Clayton and St. Paul's, part of St. Francis Xavier's (1856–60); St. Cyprian's, and the Unitarian in Wakefield Street (1856–7). The boom in church building in Mid-Victorian times was Australia wide. Despite a marked loss of population to the Victorian goldfields in the mid-1850s, there was a steady growth in the population of Adelaide and its suburbs, reaching about 18,500 by the 1860s and 27,000 by the early 1870s.

A marked change in construction occurred in about 1856 with the opening of the sandstone quarries around Tea Tree Gully and the increase in the use of 'freestone' for many buildings, so that a Sandstone and Bluestone Period, corresponding to Mid- to High-Victorian times is recognised to cover the three decades between 1860 and 1890. It is no coincidence that this period also corresponds to the active professional life in Adelaide of the architect Edmund Wright.

The hard sandstones were lighter in colour than bluestone, could be dressed more precisely to ashlar blocks and also carved, so that they were more flexible and more attractive, although more expensive. The majority of Adelaide's great public buildings belong to this period. The Tea Tree Gully (from 1858) and Glen Ewin (from 1863) Sandstones, possibly the best building stones in Australia, were dominant, with those from various parts of the Mt. Lofty Ranges (Mt. Lofty, Little Mt. Lofty, Stirling, Basket Range and elsewhere) of less importance.

The Glen Ewin Sandstone was much favoured by the architect Edmund Wright and appears in most major buildings such as the Treasury (1858), Adelaide Town Hall (1863–6), the Adelaide G.P.O. (1867) and the Supreme Court (1867). The Flinders Street Baptist Church (1861–3), Pilgrim Church (1865–7) and the administrative block at the Glenside Hospital (1870) resemble each other in the use of alternating blocks or courses of light and dark Tea Tree Gully Sandstone. The use of imported limestone (Caen or Bath) in dressings and carvings at the Town Hall, Pilgrim Church, G.P.O., and the old D. & J. Fowler's Building is notable during the period 1865–8. Marble is not commonly found in Adelaide's older buildings and therefore it is of note that in 1864, Angaston Marble was used for the steps to the Adelaide Club and Macclesfield Marble for the steps to the Congregational Church, North Adelaide.

Adelaide buildings do not commonly have carved stone ornamentation, probably because of the intractability of the sandstone (and bluestone). However, there are examples in Tea Tree Gully Sandstone in the heads at the Town Hall, Supreme Court and Adelaide Gaol, and the capitals at the Flinders Street Baptist Church. The capitals, heads and friezes at Edmund Wright House were carved by sculptor William Maxwell who was brought from Scotland for the purpose; Samuel Peters came from Sydney especially for the carving of the limestone at Pilgrim Church.

Despite the dominance of sandstone, there was still significant use of bluestone during this period. The first of the typical Adelaide bluestone schools appeared in 1874 (Grote Street, Adelaide Girls' School) and the same material was used in many churches: St. Paul's in Pulteney Street in 1867 and the Lutheran Church in Flinders Street, 1872. Building of St. Peter's Cathedral which began in 1869 in various Tea Tree Gully stones was to continue for 30 years and incorporate stones as diverse as Murray Bridge Limestone, Mount Gambier dolomite, Glen Osmond and Tapleys Hill bluestones, Pyrmont (Sydney) sandstone and West Island Granite.

The last, and probably the greatest, of the sandstone buildings,

Edmund Wright House was built between 1875 and 1878. Saving this building from demolition in 1971 marked the real beginning of wide public interest in building conservation in Adelaide.

The High Victorian decade of the 1880s was one of boom, optimism and prosperity although these were reflected more in a diversification of stone types used than in the quality of the buildings. Hotels and churches were built, extended or reconstructed. Many of Adelaide's pubs (British, Kentish Arms, Huntsman, Norwood, Robin Hood, Commercial, Crown and Sceptre, Old Colonist, Bath, Ambassadors, Avoca, Seven Stars, Tattersalls and others) owe much of their present form to this time. This is true also for many churches (Clayton, Trinity, Knightsbridge and Norwood Baptist, Wesley and others). Perhaps most important as a major element in the character of the city was the appearance of the great, two-storey bluestone and sandstone houses of Adelaide, North Adelaide and Walkerville.

New stones appeared in this period: Kapunda Marble and West Island Granite (Parliament House, 1883), Victor Harbour Granite (Bank of Australasia, 1885, and the Commercial Bank, King William Street, 1888), Manoora Sandstone (Glenelg Institute, 1877 and the warehouse in Wyatt Street, 1878), Melbourne bluestone (basalt) as plinths or foundations (Government Offices, Victoria Square, 1881 and National Mutual Life Association, Victoria Square, 1888), Sydney sandstones (Jervois Wing of the State Library, 1879–84, Mitchell Building, 1879–81, and the M.L.A., Victoria Square), and possibly Tasmanian sandstone in the Government Offices in Victoria Square. The import of stone from New South Wales and elsewhere was resented by the local producers and led to a public meeting and formal protest in 1880.

Although domestic building was to utilize stone for the next 50 years in the typical freestone veneer front and brick side walls of the suburban bungalows and villas, there was to be little further major building in stone. The 1890s (Late Victorian) was a period of great change which began with a world-wide depression. Some larger houses continued to be built in Adelaide but few major buildings belong to this or the following Federation Period. There was a change from the more-expensive traditional stones to limestone so that a Limestone Period can be recognised (1888–1950), coinciding with the architectural period, recognised by Freeland as Federation, Transition and Early Modern.

Although the Mount Gambier limestone had been used extensively in the South-East for over a century it was popular in Adelaide only for a brief period in the 1950s. Limestones from Waikerie and Ramco were used in several large bank buildings in the 1930s and 1940s (S.A. Savings Bank and the Bank of New South Wales in King William Street).

Following the Limestone Period and overlapping it was the final Inter-

national Period when stones from abroad began to assume prominence. There had been some diversification of stone types during the boom of High Victorian times but the appearance of foreign stones was not marked until well into the twentieth century. Local stone is represented by the Monarto Granite in Electra House (1901) and red Murray Bridge Granite in the plinth of the South African War Memorial, in the Adelaide Railway Station (1927) and in the Savings Bank (King William Street). The last significant use of Glen Osmond Bluestone appears to have been in the Church of Christ, Grote Street in 1925. Sandstone from Sydney produced the typically smooth ashlar walls of the Savings Bank in Currie Street (1902–04), Elders-Goldsborough-Mort (1935), the C.B.A. building (1936) in King William Street, and Elder House (1937–9).

Since the Second World War, the use of stone has been declining and in modern buildings, stone is used as a thin veneer, mainly on visible, lower parts. These buildings, clad in granite or marble from Australia, Italy, Sweden, Greece, South Africa and elsewhere are indistinguishable from those in any other city. Synthetics have been increasingly used: the artificial granite ('Benedict stone') of the C.M.L. buildings (1934–6) at the corner of Hindley and King William Streets repeated its use by that Company in other Australian cities. The extension to Parliament House (1936–9) also used an artificial granite as the plinth. Cement render (stucco) resembling sandstone ashlar was used extensively in the late 1920s (Adelaide Railway Station, Moore's (now the Law Courts), Shell House and the Liberal Club Building).

The Imperial (black) granite quarry at Sedan was opened in 1958, with the Sienna (brown) granite and the Kingston (blue) granite quarries in 1965. The Reserve Bank (1964) was clad in Wombeyan (N.S.W.) marble, and granite from Sedan (Black Imperial), Harcourt (Victoria) and West Australia. The adjacent State Administration Centre (1964) was clad in white Angaston and Paris Creek 'Sea Wave' marbles, and Black Imperial granite. The striking marble in the foyer of the Adelaide Town Hall is also 'Sea Wave'.

Out in the suburbs, sandstone from Basket Range, Manoora, Kapunda, Macclesfield and Brinkworth continues to be used for a few domestic dwellings (mainly as the tuck-pointed, variegated ashlar, unique to Adelaide) and Adelaide still retains some individuality as stone is rarely so used in the other cities of Australia.

There were about 9,000 buildings in the city of Adelaide (excluding the suburbs) in 1879. This has decreased to 7,560 at present, of which less than 5% (dated mainly between 1864 and 1884) are scheduled for protection in the Adelaide City Council Heritage Register. The building stock of Adelaide is a valuable source of social history and of all elements of the building fabric, it is stone which imparts a distinct character to Adelaide.
(*Summer 1983*)

Supreme Court, Adelaide, entrance gates and north façade. (Photo: National Trust of S.A.)

Australia in Trust

Lattice panels, doors and elaborate fretted brackets to an 1880s house. (Photo: T. Conway)

The Queensland house

Richard Allom

Richard Allom is an architect whose practice is concerned chiefly with the historic built environment. Although his work is in all the eastern states and the Northern Territory, from his Brisbane office he has developed a special interest and expertise in the care and conservation of timber buildings, and in particular the Queensland house.

Although popular legend has it that Australia's exploration and settlement was generated by the need for penal establishments, in truth the settlement of the northern part of the country was, like the southern colonies, related largely to the need for commercial and defensive establishments along the coast. More importantly, however, northern Australia was explored and settled as an outpost of the southern colonies by virtue of its mineral wealth and pastoral potential.

Here were few Government towns or military garrisons. Rather the towns and settlements that grew in the north were ports to export the produce of the vast pastoral holdings that were taken up in what is now Queensland, the Northern Territory and the north of Western Australia. Inland settlements were based around hotels, stores and later, customs posts, particularly where population due to gold discoveries demanded such facilities. It was, then as now, a frontier environment based largely on primary production.

Fortunes were to be made and the early settlers in the rush to be rich knew where their priorities lay. The production of wealth took precedence over the provision of comfortable lifestyles and in many instances it was to be one or two generations before a conscious effort in building elaborate or even moderately comfortable homesteads and housing took place.

There was a remoteness, even within a remote land, from those things that society deemed proper, fashionable, or even desirable. The earliest settlers, and indeed all those until the railways linked some of the coastal settlements with the inland, were obliged to be totally self-reliant in material goods as well as in less tangible matters such as fashion, style or good taste. Even with the advent of the railways the vast distances of northern Australia made only minor improvement to the cost and difficulty of travel and transportation of imported goods and ideas.

Finally, the climate in northern Australia bore little relationship to anything that first generation immigrants had experienced before. No familiar green valleys and winter rain greeted the early settlers, but rather tropical forests and arid desert landscapes with extremes of climate manifesting itself alternately in crippling floods and droughts, and on the coast, in heat, flies and mosquitoes, and in soaking tropical rain for weeks on end. Small wonder then that the population of this part of Australia developed a style of life that was somewhat different from their southern counterparts and developed a form of housing to suit that lifestyle and the climate in which they found themselves.

Throughout northern Australia architectural forms developed in direct reponse to climate and environment and, because of the isolation, paying

Australia in Trust

relatively little acknowledgement to prevailing fashion or architectural style — the perfect conditions for the development of a true vernacular style.

In the southern parts of the continent models and patterns brought from Europe formed the basis for architectural style and these, working satisfactorily in a climatic and social sense, have formed the basis there even for modern architectural thinking. In the north those models, even in the early centres of growth such as Brisbane, Townsville or Charters Towers, generally prevailed only until the 1860s and these were even then severely modified to suit local conditions, by the use of verandahs, roof ventilators, shading devices etc. In the more remote and less populated centres European models were even less of an influence and the use of local materials fashioned into a style of housing in direct response to climatic and functional influence was common.

Far left Detail of building on page 218. Notice that the middle post is integral with the frame while the others simply support the bottom plate.

Top left A high style house showing use of corrugated iron and applied timber decoration to the verandahs. (Photo: T. Conway)

Bottom left Axonometric drawing of a late (1920s) single skin Queensland house.

Above An early light framed house, with timber shingle roof, showing exposed frame and the purely functional bracing, i.e. non-decorative. Although the main frame of this house is sawn timber, the sub-floor structure is hand worked and massive.

By the early 1870s technical developments within the building industry — particularly the general availability of sawn timber and corrugated galvanized iron — acted as the necessary catalyst to this ferment of vernacular potential and, almost spontaneously, the distinctive 'Queensland House' resulted. That is, the characteristic light timber framed, pyramidal iron roofed house on high stumps with verandahs, sun hoods, lattice screens and roof ventilators, and other functional devices integrated with the whole in a cohesive and recognisable vernacular style.

Much work has been done in recent years in an effort to accurately pin-point the origins of the style. The results to date have been inconclusive. Certainly, the ready availability of good timber such as Hoop and Bunya Pine is an important consideration in the development of a timber vernacular. A recent and thorough report to the National Trust of Queensland on the nature and evolution of the Queensland House by Donald Watson, suggests that the style is related, at least in part, to the work of architects Richard Suter and Benjamin Backhouse who, with an interest at that time in Medieval building and Gothic architecture, designed a series of 'half timbered' buildings showing exposed stud work. Watson also points out that through the development of this style as schools and schoolmasters' residences for the government, the tradition was spread throughout the colony.

Another school of thought points to the bamboo and cane housing of some of the south-sea islands and draws a comparison to the high-set light framing and necessary cross-bracing exposed on the outer face of the external walls to that of the Queensland house.

Still another argues that the setting of the Queensland house high above the ground was in direct response to the problems of white ant

Australia in Trust

Top Six metre stumps and battening at the rear of this Brisbane house. (Photo: T. Conway)

Left The tradition of moving houses is a long one, as this early photograph from the family album of Brisbane house removalist, Roy McKenzie, shows.

attack in this part of the country and that the logical reaction was to lift traditional bedlog and slab construction up on stumps where insect attack could be detected before causing damage to the main or structural frame and indeed documentary evidence supports this theory.

Certainly, the development of more sophisticated sawmilling techniques meant that timber could be cut with some accuracy and precision, leading to generally lighter sections. The introduction of corrugated galvanized iron to Australia meant not only a lightweight and waterproof roof requiring lighter roof frames but the inherent structural quality of the iron itself was soon recognised: that is, curving the iron by rolling it through a mill made verandah roofs virtually self-supporting, requiring almost no structural support. Some believe that this dependence upon sawn timber and corrugated galvanized iron was imported from North America via Californian gold prospectors, and only superficially modified to make the distinctive style so well known in Queensland today.

As to the plan form of the Queensland house, one school of thought suggests that the Indian bungalow has a place in its development and certainly this seems not unlikely. The European settlers of tropical northern Australia would have had little experience of living in tropical conditions and might logically have looked to British experience in colonial India. Certainly the square plan form with verandahs on four sides with cooking and ablution facilities in separate buildings tend to support this theory.

There is some evidence to support all these theories and some surviving housing of the period 1850 to 1870 occasionally throws interesting light on the sources of the style. For example, in parts of Queensland and northern New South Wales early housing survives in which the stumps and the structural frames (particularly the corner posts) are integral as opposed to later vernacular tradition where the stumps and the frame are separate entities.

Some of these early examples display a rich mix of slab construction and light-framed construction and the evidence deserves careful dating, recording and analysis.

Whatever its origins however, two points must be conceded. The first is that the style, as it developed to its high point within a decade of its conception, is unique to Australia, and, it has been argued, is the most important contribution that Australia has made to world architecture. The second is that the style developed spontaneously and quickly reached its zenith in a structural sense that would not soon be equalled. Architectural and planning considerations aside, the Queensland house of the period 1875 to 1890 is one of the most economic in use of material, and in engineering terms can be likened to the development since the last World War of motor cars without a chassis or 'monocoque' construction.

The walls, of lightweight construction will hardly stand alone and depend, like a house of cards, on the support of each other. Internal walls have no framing at all and boards span from floor to ceiling with only a single horizontal rail to clamp them together. Similarly, external walls of fine sectioned tongue-and-groove timber are hardly weatherproof and depend upon the protection of the wide encircling verandahs.

Yet the buildings are remarkably strong and surprisingly weatherproof. Indeed these lightweight and apparently flimsy structures have such strength and rigidity that a tradition of moving whole houses soon developed — a tradition that survives to this day. Houses were picked up and moved hundreds of miles across country, pulled by horse or bullock teams, traction engines and even floated on rafts to new locations usually as single entities, or in the case of major structures — simply cut into two or more pieces and re-connected at the new site. Their builders understood perfectly the principle of tying each piece to each other so that even in cyclonic conditions these light, flimsy structures stand firm against the winds and rain.

In less extreme conditions too these little houses perform to the advantage of their occupants. In coastal tropical Australia they provide dark and therefore, psychologically at least, cool interiors during the day with external walls totally shaded from the direct rays of the sun, the verandahs providing a cooler spot to sit as the house heats up from radiation transmitted through the roof.

In the evening as the temperature drops the light construction sheds quickly the accumulated heat of the day enabling the main body of the house to be used once again. Indeed the house elevated on high stumps catches every available breath of air even beneath the floor itself, adding to the cooling process. The verandahs, often further protected from the sun and rain by lattice or louvre screens, constitute an outdoor living room, particularly in wet weather and generations of Queensland housewives have dried their washing during the monsoonal months on makeshift clothes lines strung between the bearers under the house.

In other ways the Queensland house was conceived and developed with an acute awareness of the broader environment. In the summer months

Australia in Trust

This early slab building shows a transition from bed-log construction to stump and bearer construction.

the glare of the sky in northern Australia is extreme and demands some relief for the inhabitants of that part of the country. The use of lattice, louvres and fretwork to the verandahs has the effect from within the house of breaking down that glare while still allowing the free passage of air into the house. Indeed the tradition of fretwork usually carries through into the house proper with intricately carved fanlights above doors to allow ventilation into those rooms.

The effect of louvres, lattice and fretwork to the external architectural treatment of the house is no less important. In northern Australia the contrast between sun-lit and shaded material is most obvious leading to an effect that creates a wonderful pattern of light and shade. This effect is often carried down to batten screens beneath the house and between the high stumps.

These 'Queensland Houses' were built in their thousands from the northern sub-tropical parts of New South Wales right up the Queensland coast into the Northern Territory. Indeed like many vernacular styles they became an accepted language of architecture and found their way to Central Queensland and the dry northern interior where their basic advantage in a climatic sense was less appropriate. Modifications to the style were made to accommodate this difference but the house remains basically one rooted in tropical coastal Australia.

Some of the regional differences developed are interesting in their own right. For example in Charters Towers a tradition developed of creating a second shading device on to the verandah; in Maryborough the timber stump — round in most parts of the State — was roughly dressed into a square; in parts of Western Queensland the shortage of timber led to the use of fine corrugated iron as external cladding and indeed in some areas where conditions demanded it external walls were constructed as a cavity and filled with charcoal to provide more appropriate protection against the heat.

By the late 1880s the wealth associated with the mineral and pastoral boom had caused the basic vernacular structure to change somewhat and what had begun as a most economic and even spartan structure began to change in plan and in decoration. Plan forms became more complicated as larger and more substantial houses adopted the style. Consequently roof forms moved away from the simple pyramid to hips, gables and turrets. Decoration too became more elaborate and some late nineteenth century timber houses in the wealthier areas equalled any of those in Sydney and Melbourne in applied decoration — the difference being that in northern Australia the art of fretworked timber predominated rather than plaster or render as in the south. Decoration too was applied to sun hoods and roof finials in carefully worked zinc anneal.

Despite the economic slump of the 1890s and the general rise in cost of building following the first World War, the vernacular continued into this period with some modifications. Decoration became less elaborate, and indeed somewhat cruder, coinciding with the heavy decorative style of the Federation period. Verandahs were reduced in scale and extent and with their demise went another intrinsic element of the true vernacular, necessitating more substantial exterior cladding to the exposed walls in order to throw the water way. The popularity of terracotta tiled roofs at this time meant that the traditional light frame needed strengthening and gradually the Queensland house became more like its timber counterparts in other parts of the country.

It remained however in basic form as a house on high stumps, and even today in an era when concrete-tile, brick veneer and plasterboard lining is the norm for speculative housing, the Queensland house remains high-set and identifiably of the vernacular tradition. Economic considerations today demand that under the house be more efficiently used and generally it accommodates motor cars, a laundry and a space for billiard or table tennis table.

The verandah has shrunk to a tiny vestige of its former self and might more properly be called a porch, but the origins of the style are clear — the Queensland House lives. (*Winter 1982*)

Palma Rosa, Brisbane

Peter Marquis-Kyle

Peter Marquis-Kyle is a Brisbane architect who has specialized in the conservation of historic buildings.

It is a most characteristic feature of old Brisbane that the best and the grandest of its Victorian houses are situated on the tops of hills. Brisbane's meandering river and hilly ground offered an abundance of sites with prospects of water and countryside and, in this town of free enterprise, the most commercially successful citizens rewarded themselves by building their mansions and villas on elevated tracts of scrub, where the rush and toil of the docks and warehouses could be admired with detached comfort. Honest trade (or shrewd speculation) afforded the wherewithal for a stylish life and, of course, a stylish house. In the 1870s and 1880s Brisbane's population and trade were growing fast. New houses were being put up on the hillslopes near the river and town like a diagram of who lived where — wooden cottages where work was done, villas on the hills, and middle ranking houses in between.

To those who arrived in Brisbane by steamer late last century this impression of the houses of a leisured class squatting on the hills and ridges must have been striking. Even now, when the growth of the city has spread the suburbs in an even smear out over the ground, the chimneys and gables on the inner suburban hilltops still evoke that former impression.

Andrea Giovanni Stombuco arrived in Brisbane in about 1875 — a bright time for a practical and energetic fellow with more than a bit of ambition. Born in Florence around 1820, Stombuco travelled widely in Europe and South America before settling in Cape Town where he married. For some years he operated a quarry before setting off for Victoria with his wife and infant child in 1851. It may be assumed he landed on his feet in goldrush Victoria and he worked as sculptor, mason and building contractor. By the late 1860s, Stombuco was at work as an architect, an occupation in which he was probably self-taught, and an occupation he kept up until his death in 1907. He settled and worked in various places in Victoria and New South Wales before coming to Brisbane. Late in life, he moved alone to Perth.[1]

Stombuco's architecture was competent, workmanlike, and diverse in style. His surviving work shows a capacity for both the austere and the flamboyant. In his respectable output, Roman Catholic churches predominate, but there were schools, institutions, commercial buildings and houses as well. From his fifteen years work in Brisbane, at least twenty buildings still stand. He is best known now as the designer of the flamboyant Opera House (later Her Majesty's Theatre but now demolished) in Queen Street, Brisbane, a building always popular for its elaborate facade.

As a personality, Andrea Stombuco was stylishly extravagant — a lavish entertainer, a gambler, fond of music and the good life. These attributes, along with his evident success in commerce, made him just the type to build himself a villa on the hill, and this he did. *Sans Souci*, later called *Palma Rosa* was built in 1887. Its high three-acre site offered splendid views of two reaches of the river and the city. As a piece of architecture, it nicely typified the grand villa as it occurred in Brisbane.

A newspaper account of the celebrations at its completion noted that:

Palma Rosa, Brisbane

'... The villa, which from the structure and position is undoubtedly one of the finest in or about Brisbane, is of stone from Petrie's quarry throughout, and the architecture is of the Tuscan Doric order. The tower, which faces the river immediately opposite the Toombul point is 90ft high and a view up the Bulimba Reach as well as for a considerable distance down the river is obtained...'[2] The article goes on to describe the form and finish of the house and to evoke the back-slapping speeches and sumptuous entertainment of the occasion.

Despite the description 'Tuscan Doric' we might now describe the

Right The villa with its tower seen from a nearby park.
Below The north front.

221

house as having Italianate features. The most notable architectural feature, the tower with its mansard roof, is prominent in the southern front of the house, facing the river. Otherwise the design includes the expected features of the Victorian villa, but the scale is, by Brisbane standards, remarkable. F. Lord, an early writer who has passed on a fascinating mix of fact and hearsay about Brisbane's grand houses, mentions the small size of Stombuco's family and wonders at the scale of his house.[3]

Andrea Stombuco's taste for extravagant living, shared by his wife, his son and daughter, his gambling and the necessary costs of running such a house brought the family into financial difficulties. *Palma Rosa* was sold to other owners better able to manage the expenses of the stylish life. The story is told that there were race day parties on the southern verandahs with their grandstand views of the Breakfast Creek racecourse, and that bookmakers called at the house to take the guests' bets.[4] Later, *Palma Rosa* suffered changes as its grounds were subdivided and built on and the house converted to a hospital. Its elegant verandahs were removed and most of its original decoration obliterated.

In 1971 *Palma Rosa* was bought by the Queensland Branch of the English Speaking Union for use as its headquarters and as a venue for meetings and social occasions. Work has begun to renovate and adapt the house for this new function and to reconstruct some missing parts of the fabric. The expansive verandahs, with their cast iron filigree columns, friezes, and balustrades, their curved iron roofs and their timber lattice valances, have been entirely reconstructed.

In plan, all the floors are similar, with rooms of very generous size opening off both sides of the central halls, which run the full depth of the house, finishing at the tower. Two or three rooms on each floor have projecting bays looking to south or west, and there are verandahs on all four sides of the upper floors and a loggia on the bottom floor. The house is set into its sloping site so that a dozen stone steps give access to the first floor verandah and the front door. The principal rooms are on this floor and the greatest elaboration in finish is to be found here. The hall floor is supported on a barrel vault and paved with minton tiles, the cornices and ceiling roses are quite elaborate and there is an arch supported on fluted pilasters breaking the hall in the centre. A simply decorated stair leads to the upper floor where bedrooms are found, while the kitchen and servants rooms are on the bottom floor. The cedar joinery is of the expected substantial sort, as are the chimneypieces of Sicilian marble.

Palma Rosa's varied history has left it almost bare of original furniture and decoration, its surrounding garden and outbuildings are gone, and the sandstone walls have been overpainted. But the fragments of detail which do survive, and the scale and siting of the house, can help one to imagine the villa when the Stombuco family lived there with their five pianos and '. . . dancing was indulged in by a goodly number until an early hour'.[5] (*Winter 1983*)

Palma Rosa, Brisbane

The view of the river, the city and the racecourse from the upper verandah.

The Palace Hotel, Perth

Penny Grose and R. McK. Campbell

The Palace Hotel was built for John de Baun in 1895–96. De Baun, an American, emigrated to Australia to try his luck prospecting for gold. Being unsuccessful at Ballarat, he worked as a station hand at Wilcannia and then moved to the mines at Silverton to establish his first hotel. In 1885 de Baun followed the rush to Broken Hill, where he built the Grand Hotel. He is reputed to have made a quick fortune before moving to Adelaide where he lost it in a stock exchange crash in 1892. The Coolgardie rush provided de Baun with yet another opportunity and he had considerable success with the Great Western Hotel in Coolgardie before moving to Perth in 1894.

The Palace Hotel was a magnificent three-storeyed building designed to fit in with the Perth theme of towers, domes and cast iron. It was destined to become *the* meeting place in the city. In 1901 the hotel was leased to one Mr. Glowrey, who published an advertising brochure in 1904 after spending some £5,000 on improvements and redecoration. Parts of the brochure bear quoting: 'The vestibule is a noble one, extending the full height of the building, and lighted from the roof level by stained glass windows ... (The William Street bar) is fitted with massive, polished cedar counters and fittings, with moulded panels, and Ionic pilasters, and the large bevelled mirrors, and Lincrusta Walton panels at the back.'

In 1966 Professor Freeland, in his book *The Australian Pub*, wrote: 'the epitome of the whole story of the west can be found in the Palace Hotel in St. Georges Terrace. Built ... to satisfy the taste and ambitions of the bursting colony, it is one of the few, such as the Menzies in Melbourne (now demolished) that retains the quality and atmosphere of Victorian opulence.'

This passage was much quoted in 1973–74 when the battle 'Save the Palace Hotel' was being waged. The hotel had been labelled as a development site and was bought as such by the Commonwealth Bank in 1972. The bank was adamant in its intention to demolish the hotel but the National Trust took a stand and was amply supported by the public at a meeting early in 1973. Mr. Reg Walker, then Secretary of the Australian Council of National Trusts and a member of the Committee of Inquiry into the National Estate was present as the invited chief speaker.

The Commonwealth Bank eventually agreed to defer decision until the Committee of Inquiry made its report. The Palace was saved.

In 1981 the hotel and certain adjoining buildings were sold to the Bond Corporation, which subsequently announced its intention to erect a multi-storeyed building on the site. However, the question of the Palace was no longer a matter of straight retention or demolition. During the earlier period of conflict, both the National Trust and the Heritage Commission had conceded that redevelopment could occur given the retention of '... the William Street and St. Georges Terrace facades, plus the entrance foyer and staircase, and, if possible the dining room.' This became a condition of sale to the Bond Corporation.

The Corporation commenced work on a multi-storeyed building being erected on the north-east corner of the site, the original Palace Hotel being retained in the south-west corner.

Australia in Trust

The Trust knew nothing of the plans until the Perth City Council had completed its deliberations but then a series of meetings took place at which the proposals were frankly discussed. An agreement was reached between the Trust and the Bond Corporation which goes much further than the concession quoted above. Not only will the facades be kept, but the whole of the original de Baun building. The entrance foyer, lounge, saloon and public bar, the basement bars directly below and the public rooms and bedrooms on the two floors above will be restored. The north and east bedroom wings, added by Glowrey in 1904 have been demolished.

It is thought that the architectural integrity will be maintained and enhanced and it might be possible to reverse some of the past unsympathetic alterations and additions. The magnificent dining room has been the topic of much discussion. The contractor insisted that it was in the middle of vital working space and the architects were not able to find an alternative. Thus, it was agreed that the dining room be demolished and then reconstructed at the end of the project.

At a meeting, held in September 1981, of the Trust, the project architects, the engineers and representatives of the Bond Corporation, including Mr. R. McK. Campbell (whose commission is to prepare an Historic Structures Report and to liaise with the Trust), it was agreed that the main areas of concern were:

(a) Recording, particularly of those structures to be demolished, in accordance with the guidelines established by ICOMOS Australia.
(b) Safety of the remaining building while work is in progress.

Far left Main entrance foyer.
Left and right Details of main staircase.
Above right Main column in Lounge Bar.

(c) A philosophy or approach to the conservation problems.

The third priority is still being fully developed, for example, the need for careful consideration and consultation on the redesigning and rebuilding of newly exposed rear and end walls of the old Palace and their relationship with the new building.

Mr. Campbell has made several reports to the Trust Council during 1981–82.

The following is a resume.

Survey and recording activities have clarified the pattern of alterations made over the years. Mr. Glowrey was associated with the hotel for many years and was also responsible for the 1930 alterations. New toilets were built for the bedrooms and bars, and the basement billiard room was converted to a bar.

During the 1950s more bedrooms were added and others upgraded. Those of the east wing were 'modernised' with built-in fittings, acoustic tile ceilings and flush-panelled doors which completely swamped their original character. There were also new bar fittings. Balconies and balustrading were rebuilt incorporating the original cast iron and a new granite plinth was installed at footpath level.

In the preparation of an Historic Structures Report all areas have been measured and plotted in plan, section and elevation. The facades, as a whole and in detail, have been photographed. As part of their total programme of recording the National Estate, the Australian Survey Office has carried out a photogrammetric survey of the main elevations.

For its age the structure is in sound condition, the original de Baun building being noticeably more solid than the additions. External walls are of Northcote bricks from Melbourne (these were superior quality to contemporary local bricks). The face has been painted over the years and may not be possible to strip without damaging the bricks. Wall systems show very little sign of movement, but there is some damp penetration which will need attention during restoration.

Floor construction is generally of oregon joists covered with pine boarding and there are no obvious signs of decay. Ceilings are of pressed metal in a variety of patterns with some replacement plaster sections. The quality and condition vary considerably.

Wind pressures on the intersection of St. Georges Terrace and William Street are high and variable. There is perceptible flexing of components in these conditions and this may be exaggerated by the new building. Roof structure and covering will need to be carefully considered in this regard. Traffic vibration is also severe and must have an effect on the structure in the future.

The Piling Sub-contract has now been completed, concluding Stage One of the project. All short term protection by way of new roof and wall flashings, wall and parapet stabilizers monitoring devices and protective panels to openings have now been installed and found to be adequate during Stage One work. Although there was some damage to the building fabric (mainly cracks to internal walls) this can be eliminated during future re-furbishment work and appears to pose no permanent problems. Future construction work will have less impact on the historic hotel and, in due course, will serve to stabilize the building fabric.

The conservation aspects are obviously on-going throughout the whole project; new evidence will appear which will need to be consistently fed into the conservation philosophy.

Professor Freeland's comment that the Palace Hotel represents 'the whole story of the west' takes on another meaning now — the current project represents a step forward in the living together of the old and the new, of conservation and change, and also in the input the National Trust is able to provide. The level of co-operation between all concerned is indicative of an increased public awareness of our heritage. (*Winter 1983*)

Yarloop, Western Australia

Penny Grose

Yarloop township, nestling in the foothills of the Darling Scarp, is approached through rich undulating country. Coming down from the main Perth–Bunbury road, you pass the modern Millars timber mill, hidden from the actual township. Further down, past two neat rows of identical timber houses facing each other, you cross a railway line into the main street, which leads to the old timber workshops, which still dominate the town.

There is great pride in the heritage of the town and those who have organised the restoration of the massive timber complex have the full support and co-operation of the community. A number of men employed

Yarloop, Western Australia

at the Workshops for various lengths of its history have stayed in the district and are now involved in both advising and actually restoring machinery. These men range in age from nearly 80 to their early 30s. Materials have been purchased at cost price from the local merchants; farmers have lent machinery for the vast clean up that had to be done; a local photographer has made a complete photographic record; the Shire has been active; and Millars, the company that owned and operated the Workshops, and which is vesting the land in the Shire, has been more than co-operative.

When the Perth–Bunbury railway was opened in 1893 the south-west coastal areas were prime for development especially for the timber business. The already experienced Millar brothers arrived in Bunbury in 1894 having come from Melbourne initially to participate in the gold rush, and by December of that year had selected a site for milling near Wagerup, just to the north of what was to become Yarloop. The site was chosen for its accessibility to the Jarrah forests and the closeness to the railway. The early mill on the site of the Workshops was producing mostly sleepers and paving blocks in 1895, and was attached to the main railway line by a short access line. It was the 'yard loop' of this line from which the name Yarloop was derived.

By the turn of the century the Millar brothers had established a successful business. There were three mill sites on the Mt. William Plateau, the earliest being Waterous Mill, one at Mornington near Harvey and two at Hoffman. In 1898 the company purchased the freehold of the northern section of the Yarloop site, and in 1901 the southern section was purchased. Plans were underway to consolidate the company's operations in a central place. Thus, in 1901 the transformation of Yarloop from a mill to a major workshop began.

Operations at the Workshop began to decline in the 1950s and then completely ceased in 1977. The use of rail had been phased out, logging and carting being increasingly undertaken by private trucking contractors; the fitters, boiler makers and others were no longer needed. Electric power was another technological change requiring less labour than the earlier steam and diesel power systems. By the time of the closure only basic maintenance crews were employed at the Workshops.

However, concern about Yarloop's historic importance was voiced before 1977, primarily because the Workshop's state of disrepair was snowballing, equipment was 'disappearing' and some demolition was occurring. Margaret Feilman, now Chairman of the W.A. National Trust, was the catalyst for the classification of Yarloop as a Conservation Area in 1974. A grant, supplemented by the Shire of Harvey, was obtained for a survey of the 30–hectare site. A Steering (Interim Management) Com-

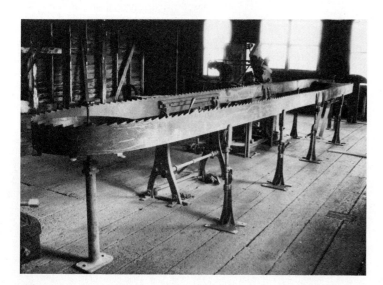

Yarloop, Western Australia

Above Main complex and Truck Shop (before restoration).

Left Looking south from Saw Shop.

Far left Restored band saw.

mittee was formed in 1979 comprising representatives of the Museum, the W.A. Heritage Committee, Murdoch University, the Department of Tourism, and Millars (W.A.) Pty Ltd. Mr. Geoff Cattach, now the enthusiastic Chairman of the Restoration Committee, was the representative from the Shire of Harvey. The Committee's brief was to make a recommendation on whether to restore or not. The decision was positive and work began in July 1981. The comprehensive survey deals chronologically with the major systems of the Workshop complex, the railway, haulage and power systems. Each is a fascinating record of technological change.

The railway system was vital to Millars operation, and therefore the maintenance of locomotives was of prime importance. In the 1930s the company owned nearly 400 kilometres of railway line and 25 steam locomotives. The Workshops central complex, developed in the first decade, contained three main areas, the Loco-running Shed, the Boilermakers and Blacksmiths Shops. The Loco-running Shed was built over two existing railway lines, one leading to the Fitting and Machine Shop, and the other into the Tender Yard via the Boilermakers Shop. Two engine pits ran the length of the structure. In the blacksmithing and boilermaking areas, springs and tubes were manufactured or replaced.

Repair and manufacture of moving parts was carried out in the nearby Fitting and Machine Shop. The shop contained lathes, planing, shaping and drilling machines. There were also facilities for casting iron and brass parts, the moulds being derived from wooden patterns made by a master pattern maker. Most of the patterns are still in the Pattern Store, and Yarloop's patterns had an excellent reputation, often being requested for loan by other foundries in the State. The vertical furnace of the Moulding Shop is still mounted on the west wall; the iron bucket, wooden box and iron poker remain.

The Plate Working Area was created as an extension of the Fitting and Machine Shop. To the south of this was a large clock mounted in a glass-fronted box, standing 2.5 metres from the ground. The actual clock was made in London and has an eighteenth century face. It was sold in 1979 but was returned and has been completely restored by Mr. Walter Cattach.

One of the major operations carried out at the complex was the construction of rolling stock, and a Truck Shop was built in the early 1900s. Wheels and shafts were originally turned on lathes in the main workshop, the shafts being then cut to length by a mechanical hack-saw located in the Tender Yard. After 1918 the wheel lathe was moved from the Fitting Shop, together with a grinder and a hydraulic press, to the Chaff Shed in the Tender Yard.

Thus, a vital function of the Workshops was the maintenance of steam locomotives and trucks. Indeed, the last steam locomotive in regular service in Western Australia was Millars 'G'-class 4–6–0 engine No. 71, the 'Menzies', which ceased operation in 1973. The engine was given by Millars to the Hotham Valley Railway, and the Restoration Committee would like to retrieve it or to possibly use it on a 'marriage basis' with the tourist railway. When the Workshops are open to the public, the concept is that they will be a working slice of history, and the 'Menzies' could form the linch-pin of such an operation.

A second major system recorded in the industrial archaeology survey is the haulage system. The timber industry was dependent on the means of transporting felled timbers to a railhead. Stables were built west of the main complex for the Workshop horses, whose job was to cart wood and machinery around the yard and town. A similar stable arrangement was to be found to the north-east, adjacent to a large paddock in which the mill horses rested before being returned to work. Feed for all horses (and Millars had up to 1,200 of them), was supplied from a Chaff Store located in the Tender Yard. This was replaced in 1920 with a larger building located to the south of the Truck Shop. There was also a Saddle Shop for making harnesses and other horse trappings. The use of horse and steam was gradually phased out in the early 1930s, being replaced by tracked vehicles and trucks. This naturally resulted in functional changes for the Chaff Store, saddlers and stables (the Saddle Shop was converted to a library), and in the addition of other buildings.

The restorers are particularly proud of the complete operational restoration of the Saw Shop, nearly all parts having been retained. The first types of saw to be restored were band-saws, circular-saws and cross-cut-saws. It is thought that Millars erected the only horizontal band-saw in Australia at Yarloop.

There was also a range of auxiliary services supplied at Yarloop. Firstly, various storage buildings, including a bulk oil store, were needed. The Mill Store was important, containing small pieces of equipment ready for dispatch to mills when required. Some of the items included loco fittings, grindstones, welding rods, rivets, bolts, pipe fittings, gasket rings and nails.

Millars was also a major trading company, necessitated by the fact that many of the mill sites were in isolated areas and the company needed to service them with the everyday requirements of a community. Yarloop had a general store, carrying groceries, clothing and general merchandise. There was also a bakery which supplied fresh bread daily to the mills, and a large butcher's shop. Meat was obtained from a slaughter house located to the north of the Workshops. (If someone died in the early days, the butcher's cart, used to bring meat up from the slaughter house, was scrubbed out and used as a hearse!)

The actual restoration began in July 1981 and proceeded with great enthusiasm. Much jarrah planking had to be replaced, and in some areas, such as the Loco-running Shed, dressed jarrah had to be obtained. The Restoration Committee was established as a working group — one of the members, who worked in the Workshops as a boilermaker's assistant for 20 years, has restored 175 window frames himself. John Sterritt, the last saw doctor, has done most of the restoration on machinery in the Saw Shop. Ray Dixon, one of the last plumbers, has supplied detail on plumbing operations. There are many others who willingly gave of their time and knowledge to the project. It is planned to open the Workshops to the public in 1984, to coincide with Millars centenary celebrations. However, it is probable that the entire project will take at least until the end of the decade.

The Yarloop Workshops were basically the township itself, in the sense that life centred around them. Now the people of that dwindling town are working together towards the reconstruction of their history. It is a project with enormous social significance for the town, the Shire and the whole South-West. If Yarloop was once the mecca of the timber industry, it can become so again in several ways. Firstly, it can be the historic link for mills and mill towns that have closed and are literally disappearing. All mills in the district were serviced from Yarloop and there is thus a genuine link if remnants of equipment are brought from other places to Yarloop for retention and display. Secondly, once sections are operational, Yarloop has great tourist potential as a working model of the industry and its social history. (*Summer 1982*)

Historic bridges of Australia

Colin O'Connor

Colin O'Connor is Professor of Civil Engineering at the University of Queensland.

Above The railway bridge at Noojee, Victoria, is a fine example of timber trestle construction. Now out of service, it replaced an earlier bridge built in 1919 and destroyed by fire in 1939.

Above right The Djerriwarrh Creek bridge, opened in 1859, is now out of service. It is immediately adjacent to the main highway between Melton and Bacchus Marsh, Victoria.

Right The Richmond Bridge, Tasmania, is generally regarded as Australia's oldest surviving bridge. It was opened in 1825.

During 1981 and 1982, the author has carried out research on historic bridges of Australia with the support of a grant from the Australian Heritage Commission. The grant arose from discussion between the Commission and Professor R. Whitmore, Chairman of the National Panel on the Engineering Heritage of the Institution of Engineers, Australia. The work is being carried out under the general supervision of both of these bodies, and is the first co-operative endeavour of its type in engineering in Australia.

The initial aim of the project is to produce a booklet with the title 'How to Look at Bridges'. This will provide the public and others with the necessary information to recognise and describe bridges of historical importance. It will contain illustrations of Australian bridges chosen so as to provide the background against which other bridges may be judged.

As part of the project, the author is producing the first draft of a register of Australian historic bridges. This has required visits to the road and rail authorities in all Australian states and has permitted the author to visit and photograph some 300 bridges. The work is now nearing completion, the only outstanding tasks being in Western Australia, and the completion of work in Queensland.

A bridge is not only a technical achievement, but is also of social importance. Civilization requires transport, and this in turn requires the construction of bridges. It is not surprising, therefore, that a study of historic bridges takes one to many of the historical places and routes of Australia.

There is little doubt that the earliest substantial Australian bridge crossed the Tank Stream in Sydney. Following an earlier timber bridge (possibly completed about 1789), a stone bridge was built about 1804. Unfortunately, it was of poor quality and needed reconstruction in 1806 and 1811 (O'Grady, 1981). Early maps — for example that by M. Lesueur, 1802 — show other bridges in the Sydney area, presumably of timber, (see *The Roadmakers*, p. 9) and there was a timber bridge at Parramatta in 1794. None of these bridges can be seen today. The shoreline of Sydney Cove has been greatly altered, and the site of the old Tank Stream bridge now lies about the intersection of Bridge and Pitt Streets.

The Richmond bridge in Tasmania is commonly classed as the oldest existing Australian bridge. It crosses the Coal River on the route from Hobart to Port Arthur. A stone on the parapet is engraved with the date, 1823. The first stone was laid in December 1823, and the bridge was opened to traffic in January 1825. Structural defects became apparent early in the life of the bridge, and major repairs were effected in 1884, including the construction of new stone cutwaters around the central three piers.

The Hobart Rivulet runs through Hobart town, and is crossed in turn by Molle, Barrack, Harrington, Elizabeth and Argyle Streets. Early masonry bridges were constructed at each of these locations, the oldest being the Wellington bridge (1816), at Elizabeth Street. More recently,

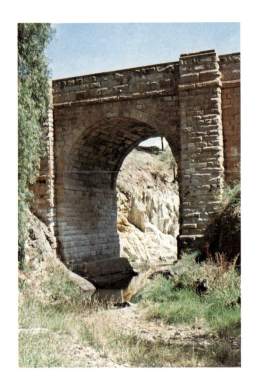

the rivulet has been completely roofed in the city area, but if access is made through this drain, the underside of the old Wellington bridge may be seen. Its location is recorded by a plaque set into the pavement of Elizabeth Street near the Cat and Fiddle Arcade.

Tasmania has at least ten other convict built bridges (Smith, 1969). The best known are the uniquely carved Ross bridge (1836) and the Red bridge at Campbelltown (1838). Some smaller bridges are at Tacky Creek (1836), Risdon (1838) and Lovely Banks (c. 1840). Less well known is the bridge at Jiggler's Creek, on the Midland highway at the southern approach into Launceston. It was authorized by Arthur in 1833, and completed in 1835. It is a single span masonry arch with approaches contained by masonry walls which extend also across the bridge. Although the span of 6.1 m is small, the archway height of 8 m, and the total length of 65.5 m make it an impressive structure.

In New South Wales, David Lennox was appointed Sub-Inspector of Roads in 1832, and Superintendent of Bridges in 1833. He built at least six bridges before transferring to Victoria in 1844 (O'Grady, 1981). His three surviving bridges in New South Wales are the Horseshoe bridge on Mitchell's Pass at Lapstone, west of Penrith (1833), the Lansdowne bridge across Prospect Creek in Sydney (1836) and the bridge in Church Street, Parramatta (1839). The Horseshoe bridge has recently been extensively repaired by the Department of Main Roads, and further work is planned for completion by its 150th anniversary in 1983.

In the Lansdowne bridge, the lower edges of the slender stone arch are bevelled to provide improved flow of the stream in flood. This feature was first used in France by Perronet at Neuilly in 1774 and then by Telford on the Over bridge near Gloucester in 1828. Lennox worked with Telford before coming to Australia, and this influence is visible in the Lansdowne bridge.

Of the bridges built by Lennox in Victoria, it appears that none remain. His greatest achievement was Princes bridge, Melbourne (1850). This had a single span of 45.7 m and a total length of 97 m. It was replaced by the present bridge in 1888, apparently not for structural deficiency, but to allow an increased waterway at the time of the development of the River Yarra.

Although Victoria has many existing bridges dating from the 1860s, the earliest of these appears to be the unpretentious Quamby bridge (1856) north of Warrnambool, on the Caramut road. Another fine early Victorian bridge crosses Djerriwarrh Creek, near Melton. Constructed about 1859, it is a single span bridge with fine proportions, and a clear height greater than its span. Unlike most other Victorian bridges, which were constructed of bluestone or basalt, it is built of sandstone, with a light attractive colour. Other old bridges are at Avenel (1859), Moorabool (1859) and Winchelsea (1868). All of these are multiple span arch bridges.

South Australia's oldest existing bridge appears to be at Inglewood, near Adelaide (1863). Two early bridges exist near Ashbourne. Jensen (1980) states, 'At Bull's Creek, two bridges were opened on the line of road by the Governor, Sir Dominic Daly in March 1866'. One of these is south of the town. To the north of the town, near the Strathalbyn turn-off is a fine structure apparently of the same age, and it is presumed that this is the second of these two bridges. There are other old bridges in the Mitcham reserve (1866), across the Sturt River in the Coromandel Valley (1866) and over the Ross Creek, Kapunda (1869).

The first railway lines in each state, and their opening dates are: Victoria — Flinders Street–Port Melbourne, September 1854. New South Wales — Sydney–Parramatta, September 1855. South Australia — Goolwa–Port Elliot, May 1854. Queensland — Ipswich–Grandchester, July 1865. Tasmania — Launceston–Deloraine, February 1871. Western Australia — Geraldton–Northampton, July 1879.

The first railway bridge in New South Wales was a masonry viaduct between Lewisham and Petersham (1855), replaced by a Whipple truss in 1886 (Fraser, 1981). Some early masonry viaducts still exist — such as at Picton (1867), as part of the great Zig Zag on the western descent from the Blue Mountains (1869), and to the west of Penrith (1870).

However, the finest existing bridge is probably the metal structure at Menangle, across the Nepean River (1863). This originally had three metal spans, each of 45.7 m, and extensive approaches. The main spans were supported on two box girders, 3.8 m high, decorated by a pair of external metal ribs shaped to indicate an arch form. In 1907 it was modified by the insertion of intermediate piers to halve the spans and provide for heavier locomotives. Interestingly, two similar Australian bridges also exist. The Victoria bridge, across the Nepean at Penrith (1867), was initially built as a combined road and rail bridge, and later converted for road traffic only. It was designed by Sir John Fowler, who was partner (with Sir Benjamin Baker) in the design of the Forth Bridge in Scotland. The other bridge of this type is a road bridge at Keilor, Victoria (1868).

Near Port Phillip Bay, the Victorian countryside is relatively flat, but is characterized by deeply indented valleys such as the Maribyrnong River valley, and these provided major obstacles to railway construction. Of the Victorian railways, two were built to particularly high standards — from Melbourne to Bendigo (1862), and from Geelong to Ballarat (1862). These, and the later direct connection from Melbourne to Ballarat (1889), include some major railway viaducts. The more important are the masonry viaduct at Malmsbury (1861) and the metal viaducts at Taradale (1862), Sunbury (1862), Moorabool (1862) and Melton (1886). These metal bridges have subsequently been strengthened by additional piers, and it is noticeable that where these use mild steel, corrosion has become a bigger problem than on the older wrought iron used for the original spans.

The Melbourne to Bendigo and Geelong to Ballarat railways used masonry construction for many smaller bridges, and it is common to see bluestone arches of the finest construction built across minor watercourses and roads.

The world's first metal bridge was built at Coalbrookdale, England in 1779. Rolled sections of wrought iron date from 1853 and in steel from about 1880. Australia's earliest iron works were established in 1850 at Mittagong (Birmingham, 1979). These Fitzroy Iron Works provided wrought iron for two of Australia's earliest metal bridges — the Prince Alfred bridge at Gundagai (1865) and the Denison bridge at Bathurst (1870). Both of these bridges still stand. Other old existing bridges are at Muswellbrook (1861); Redesdale, Victoria (1868); across the Orara River, near Grafton (1873), and at Echuca (1875). The Trevallyn bridge at Launceston was constructed in two halves, the first of these being built in 1864.

Reinforced concrete is a modern material. Although Smeaton (1765), Aspden (1824), Monier (1873) and others contributed to its development, true reinforced concrete beams and slabs date from Hennebique (about 1892). Against this framework it is surprising to note the comparatively early date of the Lamington bridge at Maryborough, Queensland (1896). With 11 spans of 15.2 m it is a large structure. Although designed as a series of flat arches, it would be regarded today as a continuous girder and when analysed on this basis is capable of carrying modern traffic. It is still in use, and must be one of the most significant Australian contributions to bridge construction on the world scene.

Apart from those named, there is an abundance of old metal and concrete bridges still existing in Australia, of every conceivable structural form. Full records would show that, of all materials, timber has made the greatest contribution to Australian bridge construction. The preceding sections of this paper show that existing metal bridges date from the 1860s, and that prior to that date, there exist only a few masonry structures. When particular bridges are studied it is not uncommon, however, to find that they had predecessors in timber. At South Creek, Windsor, for example, there still remains an old metal truss (1880) below a new concrete bridge (1975). Prior to these, there was a floating bridge (1802), a five span timber bridge (1813) and a laminated timber arch (1853).

The Australian landscape is still graced by timber girder bridges. It is rather too easy to assume that the earliest bridges were identical in form, or that their technology was totally imported. In fact, the latter assumption is clearly incorrect, for Australian timbers differ greatly in their physical properties from those in Europe or North America. There is room, therefore, for research on the development of the Australian timber bridge.

It is appropriate to conclude with a few remarks on conservation. Australian historic bridges are surprising in their range and variety, and are well worthy of preservation. This article has mentioned only a few and has omitted many later bridges, such as the Sydney Harbour bridge, that are clearly important. The best way of preserving a bridge is to keep it in service. A bridge requires regular maintenance if it is to be truly preserved, and the associated costs can rarely be supplied by preservation trusts. If unmaintained, a bridge not only deteriorates but can become unsightly and unsafe. Engineers are proud of their heritage. The best method of retaining historic bridges is to ensure that both they and the public are fully informed of this heritage, and that it is kept in mind when decisions are made to continue a bridge in service, or to duplicate or replace it. (*Summer 1982*)

The rock paintings of Cape York

Josephine Flood

Dr. Josephine Flood is an archaeologist who is at present a senior conservation officer at the Australian Heritage Commission. She has undertaken extensive fieldwork on Aboriginal sites in Tasmania, N.S.W., Victoria, the A.C.T. and Queensland.

1

2

4

The rock paintings of Cape York

One of the finest and largest bodies of rock paintings in the world has been discovered over the last two decades in the remote escarpments of Cape York peninsula. The existence of these rock art galleries first became known to the outside world in 1960, when the Cape York Developmental Road was being constructed. Some 320 kilometres north-west of Cairns and 12 kilometres south-east of the tiny settlement of Laura, the new road was built close to some giant slabs of sandstone, including a massive split rock, overhanging at its base. On the back wall of the rockshelter formed by the overhang, the roadworkers found large, colourful Aboriginal paintings.

News of the find of these 'Split Rock' galleries reached Percy Trezise, then an airline and aerial ambulance pilot who often flew over the rugged cliffs and gorges of Cape York peninsula. Realizing from the quantity and quality of the painting that they were likely to be only a small portion of

1 A decorated rockshelter near the Little Kennedy River. (Photo: K. Hueneke)

2 Stencil art at the Dilly Bag Shelter in the Koolburra Plateau, discovered in 1982.

3 The Quinkan Reserve, near Laura, north Queensland. At the base of the sandstone cliffs are many decorated rock shelters, including the Magnificent Gallery.

4 Giant Wallaroo, Jowalbinna. Several giant wallaroos adorn the walls of this large rock shelter.

5 Echidna Shelter, Koolburra Plateau. This huge shelter contains paintings, weathered engravings, and stone tools from several thousand years of use as a campsite.

6 Paintings from the European contact period at Giant Horse Gallery near Laura, where the main figures are a man shown falling off a horse, still holding the reins.

7 An Imjim Quinkan spirit figure at Red Bluff, near Laura. The Imjims are small fat-bellied spirits with stone axes on their elbows and long knobbly tails.

8 The Early Man site has its whole back wall covered with ancient engravings. Excavation of the shelter floor (by Dr Andrée Rosenfeld of the Australian National University) revealed that people have been camping there, making stone tools and pecking designs on the sandstone walls, for over 13,500 years. (Photos: J. Flood)

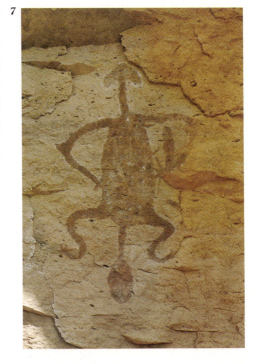

Australia in Trust

Above The central part of the frieze in Magnificent Gallery, exhibiting a high degree of over-painting.

Right An ancestral figure, more than lifesize, in Magnificent Gallery. (Photos: J. Flood)

The rock paintings of Cape York

a great body of traditional art, he embarked on a programme of exploration and recording which is still continuing today. During the last two decades several hundred painting galleries have been put on record, and it is estimated that only 70 per cent of the sandstone belt has been explored for sites.

The great belt of Cretaceous sandstone plateaux extends from the coast near Cooktown to the Hahn River in the west, and from the Deighton River to the Palmer River from north to south. The number of decorated rockshelters in this 10,000 square kilometre area must run into many thousands. In such rugged country the precise number will never be known but my recent intensive survey of an area of only 25 square kilometres produced 175 sites.

Art sites are found in rockshelters eroded out of escarpments or under the overhanging sides of massive boulders which have tumbled down from the cliffs. Many of these shelters contain only a few hand stencils, but others are major galleries with dozens, even hundreds, of figures painted on the rock walls.

The paints used were natural pigments such as ochre ground up to powder and either mixed with water and possibly a fixative, or occasionally used dry like a crayon directly onto the rock. The remains of a paintbrush found in one rockshelter had a wooden handle grooved and hollowed at one end to hold a bunch of fibre, hair or feathers. Colours are vivid, and range from purple through every shade of red, brown and yellow to white. The white is obtained from white pipeclay found in many of the creek beds.

It is not possible to obtain absolute dates for the paintings, but a preliminary chronology of rock art in the Laura region has been set up after a careful study of superimpositions, where one figure has been placed on top of another figure. From such analysis it is clear that the technique of rock engraving was used before rock painting, for in several sites paintings overlie extremely weathered engravings. When fragments of a rock painting fall from the wall of a rockshelter and get buried in the earth floor, the paint rapidly disappears, but fortunately engravings in the same situation are preserved. This has happened in the Early Man shelter near Laura, where a slab of engraved rock lay in the shelter floor covered over by stone tools, camping debris and charcoal, which was radiocarbon dated to 13,500 years old. The engravings must have been executed before, possibly long before, that campfire was lit in the shelter thirteen and a half thousand years ago.

The Early Man shelter in north Queensland is the earliest securely dated rock art in Australia although there is strong circumstantial evidence that wall markings in Koonalda Cave on the Nullarbor Plain in South Australia are more than 20,000 years old.

The main elements of the oldest engravings pecked on the wall of Early Man site are small circular hollows or 'pits', mazes, rings, and 'trident' forms which are probably bird tracks. These also occur at some of the other engraved shelters, and later other figurative motifs appear, such as the tracks of kangaroos or wallabies, human hands and footprints, human and animal figures and artefacts such as boomerangs.

When the art of painting began we do not know, but there was not a simple change-over from engraving to painting, for in a few sites engraved tracks are superimposed over paintings. Likewise we know that ochre was used from the beginning of occupation in excavated sites such as the Early Man and Green Ant rockshelters. What we do not know is who ground up the ochre pigment, and why. Was it women decorating a wooden carrying dish, men painting their bodies ready for a ceremonial dance, or ice age rock artists preparing their paints?

The sequence of techniques and styles in broad outline is from engraving to painting and from symbolized, non-figurative motifs to more naturalistic figures. Within the body of paintings there seems to be a development from simple stencils and outline pictographs to figures painted in silhouette, first in one and then in two or more colours. The more recent Cape York paintings are distinguished by their large scale; many of both the human and animal figures are more than life-size. Some of these paintings can be dated by their subject matter, which includes European motifs such as horses and guns. Apart from the first few explorers such as the ill-fated Kennedy expedition of 1848, the area remained undisturbed until the Palmer River gold rush in 1873. This was one of the most violent invasions of tribal lands in Australian history. For a decade tens of thousands of miners poured into north Queensland, with indiscriminate slaughter on both sides. But Aboriginal sorcery paintings were no match for rifles, and their traditional way of life vanished for ever.

Some rock paintings were still being done up till the 1920s, but they seem crude monochrome daubs in contrast with the fine draughtsmanship of earlier paintings, and represent the last gasp of a dying culture, the death blow being the influenza epidemic of 1922. By the 1960s, there were only a few tribal elders in the Laura district who retained sufficient traditional mythology and lore to throw light on the meaning of the paintings. They were able to point out figures in the galleries painted for love-magic, sorcery, mortuary ritual or hunting purposes. They recounted the myths of *Goorialla* the Rainbow Serpent, *Gaiya* the Giant Devil Dingo and *Bulinmore* the Echidna, and identified sacred sites bearing figures of ancestral beings.

The knowledge these old men passed on before they died has helped us to understand the meaning of some of the paintings, and to distinguish sacred from secular sites. Most of the art in fact is secular. Stone tools and camping debris are found in such shelters below the painted walls. Many of the animal and human figures are small and crudely drawn, and stencils occur of women's items such as dilly bags, of children's hands and even of babies' feet.

Sacred sites are distinguished by their general lack of camping debris, and the type of figures on the wall. These are usually large, finely executed, carefully decorated culture heroes and totemic ancestors, some human, some animal and some half and half.

The subject matter of the art is dominated by human figures. Ancestral human beings are depicted with exotic head-dresses, rayed, branched or waisted into a tall shape said to be made from a bundle of grass. Their bodies are carefully outlined and the inner areas decorated in a different colour with spots, bars, stripes or a belt. Often the figures, both male and female, are life-size or even larger. Less elaborately painted human figures without head-dresses are thought to have been painted for love-magic when depicted with a love charm pendant around the neck and enlarged sexual characteristics, or for sorcery purposes when shown upside-down, deformed or maimed. Spirit figures are distinguished by some non-human attributes and distortions of the head, limbs or genitals. Often elbow and knee joints are enlarged, and these are said to represent stone axes.

All human figures are shown in full face rather than profile, and facial features are usually restricted to eyes, occasionally a nose but rarely a mouth. The number of fingers varies from three to seven, and feet are usually shown as if looking at a foot-print.

Animals are depicted from their most characteristic viewpoint: kangaroo and wallabies in profile but fish, crocodiles, turtles, goannas and echidnas in bird's eye view. (In contrast, in the art of Arnhem Land echidnas are painted side view). Flying foxes are shown hanging upside down from the trees and birds in silhouette, but a few paintings exist of what seem to be owls, with large round eyes in full face.

Other subjects are yams, ferns, bees' nests, bark bone-containers, stone

Quinkan Gallery, near Laura. Note the x-ray technique used to show the fishes' skeletal bones and the seeds inside the fruit. (Photo: J. Flood)

axes, spearthrowers, and many 'abstract' motifs which we cannot decipher. Visual impact is created by lateral displacement of women's breasts and the far-side legs and ears of animals. Very occasionally the x-ray technique, so widespread in Arnhem Land, is used to show the internal organs of a creature.

The outstanding characteristics of the Laura art are the vivid colours and large size of figures. While this is not a narrative art, scenes do occur, such as a woman being bitten by a snake, or those of a man falling off his horse and a pack rape at the Giant Horse gallery. We will unfortunately never know whether these scenes record actual happenings or were painted in the hope of wish-fulfilment.

Elsewhere figures are arranged in aesthetically pleasing compositions, as in a gallery discovered in 1982 near the Little Kennedy River, where a giant catfish is surrounded by prawns, fish and a turtle. The subjects of Cape York paintings tend to be static, but the degree of over-painting, wide range of bright hues and decorative detail on huge animals, ancestral spirits and anthropomorphs give this art a dynamic exuberance all its own.

Anthropomorphic 'Echidna people' are the main ancestral figures in the fifty square kilometres of the Koolburra Plateau. This sandstone massif lies at the extreme north-western corner of the region, and has a distinctive art style, with no life-size figures or European motifs but many

Left A boomerang painting in red outlined with white at the Green Ant Shelter, Koolburra Plateau, showing damage from a wasp nest (on the left) and termite tunnels.

Below left 'Echidna people' in the Echidna Dreaming gallery, Koolburra Plateau. The figures were outlined in red ochre, which has now sunk into the rock surface as a stain, and infilled with white or yellow pigment, much of which has now disappeared. (Photos: J. Flood)

anthropomorphs, hand and boomerang stencils and some ancient pecked engravings.

During the dry seasons of 1981 and 1982, I carried out an intensive study of the Koolburra region, with the support of the Australian Heritage Commission and the Australian Institute of Aboriginal Studies and the aid of Earthwatch volunteers. For the first time in the Laura region every Aboriginal site was recorded, from open air campsites to boulders decorated with a single hand stencil. Two rockshelters were also excavated to give some time-depth to the Aboriginal occupation.

What emerged was a pattern in the location, size and character of the art sites. High on the cliffs there are two major galleries of paintings: massive shelters decorated with over a hundred pictures, which seem to be painted in deliberate compositions. The motifs include strange anthropomorphs, particularly 'Echidna people' — half human and half echidna — and huge hand stencils, often high on the wall and with fingers bent back as if mutilated. These appear to belong to adult men, probably tribal elders. These galleries were not used as camping places but seem to have been sacred sites of special significance, akin to cathedrals.

Each main valley tends to contain either a 'cathedral' or a 'parish church' — a medium-size shelter or group of adjacent shelters, decorated with a few dozen stencils, engravings or painted human, animal or 'abstract' figures, and including occasional anthropomorphs or spirit figures. These sites were used for camping, for example Green Ant shelter bears on its back wall paintings, including a Quinkan spirit figure, on top of weathered engravings, but excavation revealed that people had been camping there for 6,500 years.

Smaller art sites tend to cluster in zones of influence around these major sites, which are fairly close to water and near the valley mouth. In contrast, in the upper dry reaches of creeks art sites are scarce, small and usually restricted in decoration to only two or three hand-stencils. The size of stencilled hands and feet gives a clue to their owners: the most remote, valley-head sites seem to have been marked with the 'signature' of adolescents, probably boys isolated there in the process of being initiated into tribal lore.

These simple sites contrast with the more complex art of the 'parish churches' and the glories of the 'cathedral frescoes', where Aboriginal artists have vividly enshrined their religious beliefs over thousands of years. Unfortunately, the paintings are now deteriorating, for the pigment, especially when thickly applied as was usual with white pipeclay, gradually drops off the rock surface or is washed off by tropical downpours of the wet season. This did not matter when Aboriginal artists were regularly visiting the sites, touching up or re-painting the figures, but the last rock painter is long since dead in the Laura region, and the culture heroes now gaze down on empty shelters. On some figures there are only the faintest traces of the original white and yellow pigments used to infill the body, and only the outline remains as a stain of red ochre which has sunk into the rock surface.

To conserve this vanishing heritage is a high priority. This is being approached in several ways. First is practical site conservation, aimed at preventing damage to the art from animals, people, wasps, termites and the elements, particularly water. In some shelters artificial silicone driplines have been applied to the rock to divert water from running down across the paintings, in others termite mounds have been demolished to prevent them building their tunnels across the painted surfaces. However, there is little one can do to keep wasps from constructing their mudnests on the art, or to prevent the natural flaking of the rock.

The second approach is to obtain the best possible archival record of the art. Accordingly, a full photographic and written record is being made of each site, using archivally stable film such as Kodachrome. The major sites are also being recorded by terrestrial photogrammetry, which gives a full photographic coverage and three-dimensional data from stereo-pairs of photographs. By utilizing the geometrical properties of the photographs, one can obtain accurate scale plots, contours and three-dimensional co-ordinates of both the general topography of the site and its environs and large-scale recording of details of the rock art.

At the same time, some research effort, although not enough, is being devoted to developing scientific techniques to keep the rock paintings on the walls. Unfortunately, no panaceas have yet been found either in Australia or overseas.

It would be a tragedy if this complex, prolific and spectacular part of Australia's heritage were to vanish, for the rock art of Cape York, like that of the Kimberley and Arnhem Land, is a vital part of the Aboriginal heritage, and is generally considered to be not only of national but of world heritage significance. (*Winter 1983*)

Australia in Trust

South East Queensland coal mines

Raymond Whitmore

Ray Whitmore is Professor of Mining and Metallurgical Engineering in the University of Queensland. He is Chairman of the Queensland Division of the Institution of Engineers, Australia and Chairman of the Institution's National and Divisional Committees on Engineering Heritage. He is Australia's representative on the International Committee for the Conservation of the Industrial Heritage (ICCIH) and his most recent book — Coal in Queensland: the First Fifty Years — was published in 1981 by the University of Queensland Press.

The link between Australia and coal which is resulting in the country becoming a coal producer of international importance can be traced back to the earliest days of colonisation. There can be little doubt that coal was the first mineral discovered after the settlement of New South Wales, the place being in or near the present city of Newcastle, and the year probably being 1791. Further north, Major Lockyer found coal exposed in the banks of the upper Brisbane River (a few miles from where Ipswich now stands) within a year of Oxley establishing the first settlement in the present State of Queensland in 1824.

Although coal has played an important part in the development of both New South Wales and Queensland, the question of whether the achievements of this great industry merit permanent and visual recording does not seem to have been seriously considered by the government or industry in either State. It is true that a number of accounts of the rise of the coal mining industry in New South Wales have been written but they are not comprehensive.[1,2] In the case of Queensland the research has still to be done especially for the period after 1875.[3]

Attempts to conserve some physical remains of the industry in the Hunter River Valley region, which is at the heart of one of Australia's most prolific coalfields, have resulted in nothing more than an heroic one-man effort at Freemans Waterholes and a paddock of rusting machinery at Richmond Main. Slight as this result may be, it is infinitely greater than anything which has been achieved in Queensland where the only physical recognition of coal mining is a monument recently unveiled at Redbank to commemorate the State's first industrial strike. Redbank stands near the eastern edge of the West Moreton coalfield which was the major producer of Queensland's coal from 1843 until 1966, when the Bowen Basin claimed this role. The days of production for the field are numbered and within two or three decades the only signs of this once vital local industry will be hectares of devastated countryside undergoing rehabilitation through the normal healing processes of nature, assisted in some instances by the efforts of man. Nevertheless, there is still an exceptional opportunity to record the complete life cycle of this coalfield and to conserve appropriate features for the enlightenment, education and interest of future generations. The main obstacles are the high loss rate being suffered by relevant documents and physical remains, a lack of understanding of why a conservation programme is necessary and the absence of any suggestion of the form which such a programme might take.

The prime object of establishing any industry is the production of some product or the provision of a private or communal service. Over a period of time the public may develop a great attachment to the product or service and seek to integrate it into the heritage. However, the form and character of the products of mining (or the services which it provides) are generally of little public interest except in special cases such as precious stones, gold or some mineral specimens which have an aesthetic appeal. The attraction of the industry to the public lies in its history, including the production methods used, the equipment employed, the lives of the miners themselves, the organisation of the workforce and the impact of the industry on the community. These aspects hold a great deal of interest for many people particularly in the case of coal mining.

Two important points must be made about mining practice. First, because mining by its very nature is an extractive industry, its occupation of any piece of land is inevitably temporary. The period of occupation may range from weeks to centuries but once the mineral has been extracted the reason for the existence of the mine disappears and it closes down. Consequently mining structures (in contrast to most civil or industrial works) are built without permanence or architectural merit in mind. This important trait is reflected in the character of the resulting structures.

Second, in the early stages of development of any mineral deposit the easily-mined material is naturally extracted first. In the case of the West Moreton field the coal occasionally outcropped on a creek or river bank and, by opening a mine at such a place, facilities were readily at hand for transporting the product by water. The method of working consisted simply of driving a tunnel into the outcrop and removing the coal in 'rooms' dug out to each side of it. When the labour of wheeling the coal to the surface became too great, or the ventilation was inadequate, the tunnel was abandoned and another one was driven into the outcrop further along the bank. The coal was worked entirely by hand but animal or steam power in some cases assisted in bringing the coal to the surface and controlling the level of water underground. Roof support (if any), coal chutes, wharves and buildings were constructed entirely of wood and were conceived as temporary structures. Fire, flood and natural decay ensured their complete disappearance shortly after abandonment and all that remains today on the sites of these early mines — in addition to a sprinkling of coal and possibly a small, overgrown spoil heap — are surface depressions where the overburden has collapsed into the workings. Examples which date back to the 1860s can still be seen at Moggill and Redbank.

As deeper seams were exploited it became necessary to sink vertical shafts to reach them. The additional cost incurred in sinking the shaft was recouped by mining more coal before moving on, so that the size and degree of permanence of the mines tended to grow as the depth of the workings increased. In 1870 the deepest shaft on the coalfield, at Goodna, was less than 30 m but by 1885 a depth of over 100 m had been reached at Waterstown. Wood was still the chief constructional material for shaft linings, headframes, chutes and buildings but brick emplacements began to appear for boilers or engines. The only really substantial brick structures were the coke ovens.

The production of coke in Queensland was initiated in 1869 by a short-lived requirement of the railways which led to the construction of a battery of ovens at Tivoli. At about the same time, coal mining began to follow the seams away from the rivers and it became necessary to construct tramways along which the coal or coke could be hauled to the water. Signs of the associated earthworks are still visible at Goodna, Tivoli, Waterstown and elsewhere. There was also a steadily increasing use of iron at the pits, and the site of the smithy is often marked by a brick or stone foundation and a scattering of assorted metal items.

With the coming of the railways and the construction of a coal wharf at South Brisbane there was a rapid rise in coal mining activity in the 1880s. Nevertheless, the structures erected at the collieries were still the simplest and crudest which would do the job, and machines such as winding engines or pumps were shifted from site to site for re-use.

When it was found possible to mine a number of seams over a wide area from a single shaft, the surface facilities were modified from time to time to suit the operational or market requirements, the disused facilities either being incorporated in the new structures or left to decay. The Rhondda No. 1 colliery, which is the only coal mine listed by the National Trust of Queensland, is a notable example of this process. When greater capacity was required at the colliery in the 1920s, the original winding engine was replaced by a second-hand machine originally constructed in 1899 by Walkers of Maryborough for the Gympie goldfields. A great variety of discarded coal handling arrangements surrounds the shaft; immediately adjacent to it are three successive stages of coal preparation equipment superimposed one on top of another, a living example of an industrial archaeological succession!

The spread of coal mining around Ipswich has resulted in a wave of spoil heaps and desolation as mines were worked out and operations moved on. Some of the old sites have reverted to bush which is probably not very dissimilar to the original cover, as in the Chuwar area. Some have been levelled and built on — as at Bundamba — or turned into farmland as at Rosewood, while others have been reworked as open-cuts as at New Chum, leaving giant scars whose future is still uncertain. These operations frequently sliced through unmapped old workings without, unfortunately, any attempt being made to record or study their layout while this was still possible. Further opportunities will arise, however, because coal mining in the district will rely heavily in the future on open-cut working, although its extent will clearly depend to a marked degree on public acceptability of this form of mining in an urban area.

Coal mining in West Moreton spreads over an area of some 100 square kilometres. During a period of 150 years more than 100 million tonnes of coal have been extracted, and for the last 80 years it has provided direct employment for almost 1,000 men. In another 20 years it will have gone from the area for ever. To what extent does the important and valuable human effort represented by this endeavour deserve a permanent tribute, and what form should it take?

The inherent temporariness of mining structures is a source of considerable difficulty for the conservator because he is trained to provide a structure with a high level of permanence in order to minimize the need for subsequent maintenance. One attribute of temporariness is that it encourages the frequent use of rough-and-ready bush engineering, leading to quaintness, interest and rude charm.

Unfortunately, attempts to preserve the original materials of construction and building methods are likely to result in heavy maintenance costs but if the conservator makes changes in order to increase the longevity of the structures, their authenticity is likely to be lost. Moreover if the surface buildings are conserved but the underground workings are closed, the opportunity for presenting a mine as an entity disappears and one must question whether there is much point in leaving them in their original locations especially if they are inconveniently placed for visitors. The mine is also likely to be surrounded by spoil heaps, settling ponds, and polluted streams. These could, of course, be cleaned up, landscaped and planted to provide a pleasant setting for car parks and barbecue areas, but this would only add further to the unreality of the scene. The problem is not solved by accepting the grime and unsightliness and attempting to 'freeze' the site in the state in which it last operated. For example the spoil heaps may fire spontaneously or, if thermally inactive, will revegetate quite quickly unless frequently sprayed with defoliants. If an attempt is made to keep the underground workings open, the safety requirements for maintaining them in a condition where the public can inspect them are daunting, and at a colliery such as Rhondda, the cage is so small that only 4 or 5 people could use it at any one time in order to look at them. At Chatterley Whitfield Colliery in Britain these problems appear to have been overcome, and full-size static displays of coal mining over the years have been constructed in the workings which the public can study after being lowered down the shaft in the original cage. Nevertheless, the cost of making the workings safe for the public — and keeping them so — is very considerable and the economics of the Chatterley Whitfield project would need to be carefully studied before anything similar were attempted in Queensland.

What other physical remains are left by coal mining other than caved-in areas, occasional pit-head buildings and spoil heaps? The most obvious are screening and cleaning plants, coke ovens, tramways and railways. The mechanical parts of the coal preparation facilities are usually dismantled after pit closure and the remainder quickly rust away, but remnants of the coke ovens still dot the Ipswich countryside, their construction in brick and stone making them relatively durable. The earthworks of the tramways and railways are also fairly resistant to the ravages of time, although the uninformed may not always be able to recognise them. There may also be houses or townships which were built by the colliery proprietors for their workmen and themselves. Good surviving examples exist in the Ipswich area at Tivoli, Blackstone and New Chum.

However, the heritage of an industry consists of much more than a series of physical artefacts or structures. A comprehensive record must pay equal attention to two other matters — documentation and people.

The basis of any conservation programme is the collection, stabilization, classification and ultimate analysis of the documentation associated with the item. This is particularly true of an industry and, in the case of coal mining, includes government, company and private records, books, maps, photographs, plans and correspondence. In Queensland only the official parliamentary papers and the land title deeds are reasonably comprehensive and secure. Government departmental papers which are extant are extremely patchy in quantity and quality. Some material has been microfilmed but the negatives are often of such poor quality as to negate the value of the effort. Company records are practically non-existent because most of the operations were run as small family businesses and it probably never occurred to anyone to save transaction papers. A determined effort is required to collect what is still available and put it in a form and place which is accessible for further research.

Industrial heritage material is unlikely to be explicable without an understandng of the character, experience and motivation of the people who made a living from the industry. This is particularly true of the West Moreton coalfield where commercial and technical developments tended to be subordinated to family and personal demands. Many of the mining entrepreneurs were practical men who were more proficient with the shovel than the pen and it is not always easy to discover the reasons for their actions. Their families and their original homes can still be found in the district, and recording and oral history programmes are required in order to document them.

It is now possible to examine the elements which in sum could provide a comprehensive record of coal mining in the West Moreton region. They can be summarized as follows:

1. The provision of a focal point. Because activity was spread over a wide area, some kind of focus should be provided to which interested visitors or serious researchers would be attracted. This could be an old coal mine but it could equally be a centrally-placed building in the district. This is because the interpretation of site, documentary and physical material (which is just as important as the conservation of specific mining structures) can be carried out at any convenient centre.

2. The focus should provide audio-visual and scale-model introductions to the industry in the region, supplemented by a museum and an archival section holding original and copied material for the serious research worker.

3. Embodied in the focus there should be:
(i) A lecture theatre and committee rooms for use by professional institutions associated with the industry.
(ii) A specialist library, in addition to facilities for archival research.
(iii) A commercial centre where details of Queensland coals, and information on equipment for the mining and utilization of coal could be displayed.
(iv) An outdoor area where models and layouts of underground mining operations — both historical and modern — could be constructed and opened for inspection.
(v) A conference and hospitality office.

The function of the focus, which might be termed 'The Coal Centre', would be to provide a comprehensive service to the coal industry in Queensland. It would embody museum, library, archival, research, documentary, display, learned body and conference facilities and be attractive to casual visitors, researchers, commercial delegations, professional organisations and conference organisers associated with coal.

4. From the Centre, routed tours of the various remains would be available and visitors could follow them by purchasing a descriptive booklet or a tape cassette (most cars carry tape recorders these days).

5. Each site would be clearly marked with vandal-proof signs, and presented in such a way as to encourage the visitor to use his own imagination in recreating the particular scene, even if there was nothing left on the site but some earthworks.

6. A very few substantial mining relics would be preserved on site. Items would be selected on the basis of their public appeal and ease of maintenance. A headframe, a preparation plant and a coke oven are obvious examples.

One objective of the project would be to distance the heritage work as far as possible from a dusty museum image and to present coal mining history inside the framework of a dynamic, essential, on-going, industrial activity vital to a modern community. A tribute in fact to that ubiquitous mineral — coal.

The preservation and presentation of industrial history is in its infancy in Australia, but it is a topic of absorbing interest to many people. In the Ipswich area there is a real opportunity of compiling a comprehensive record of the complete cycle of a coal mining region and presenting it in the context of the dynamic future potential of coal in Queensland. Appropriate material is, however, disappearing at a rapid rate and a serious start needs to be made in the immediate future if the result is to claim comprehensiveness and authenticity. (*Winter 1982*)

Archaeology of the Burrup Peninsula

Nicholas Green

Nicholas Green, Pre-History and Anthropology Dept., School of General Studies, Australian National University, Canberra. A.C.T.

In August 1699, William Dampier sailed the vessel *Roebuck* into the Dampier Archipelago and landed on an island he named Rosemary Island. He noted the rusty colour of the stones and smoke in the distance but 'we found no other sign of inhabitants'[1] Many other voyagers, and later, settlers were to follow Dampier's journey to these rugged islands, but it was not until 1818 that Aborigines were first observed by Phillip Parker King, who sighted them paddling log canoes, or 'marine velocipedes' near Goodwym Island.[2]

American whalers also visited the area in the early 1800s to hunt the humpback whale, but very little documentary evidence remains of this period. It was only when F. T. Gregory landed on Burrup Peninsula in 1861, that further knowledge of the area developed.[3] As a result of Gregory's expedition, Captain Jarman in the barque *Tien Tsin* sailed along the coast and settled at the mouth of the Harding River in Tien Tsin Bay (Cossack).

The township of Roebourne, inland from Cossack, was soon established and, along with sheep stations and pearling bases, represented the only European habitation in the area until 1966 when Dampier township was built to accommodate Hamersley Iron Pty Ltd. 1970 saw the establishment of the town of Karratha and the expansion of the industry into the hinterland. Today Woodside Petroleum is building a natural gas liquefaction plant on the Burrup Peninsula.

The Dampier Archipelago is situated on the north-west coast of the Pilbara region of Western Australia and is comprised of about ten large islands and a number of smaller ones. A large promontory of land, the

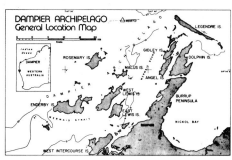

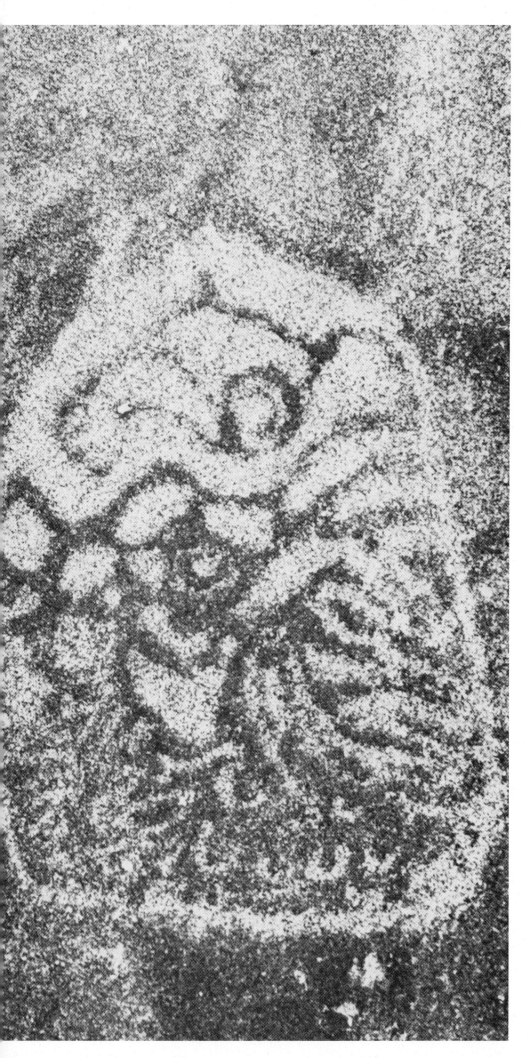

Left Infilled curvilinear design, Watering Cove.

Burrup Peninsula (named after an Englishman who was speared to death by Aborigines in the nineteenth century), forms the largest single land mass in this island group.

The Peninsula presents an ancient and striking landscape of rugged rocky terrain. These erosion resistant rocks form large outcrops and screes which have isolated soil pockets attracting sparse vegetation, primarily Tree Steppe. Drainage channels bisect the screes forming deeply eroded gullies with semi-permanent rock pools attracting a wide variety of fauna, including Spinifex Pigeons and Bar Shouldered Doves. Larger fauna such as the Euro (Macropus robustus) appear at dusk to drink at these shady pools which form an important resource today as they did in the past for Aboriginal people.

Undulating stony plains covered in Spinifex and Tussock grass extend to the sandy embayments and rocky headlands that are a feature of this landscape. Rich in marine life, the bays are flanked in areas by tidal mudflats with extensive mangrove stands. The environment pictured here is varied with rich micro-environments which are an integral part of this externally harsh landscape.

Early accounts describe the obvious friendliness of the Aborigines encountered. R. J. Sholl, the then government resident at Roebourne, wrote in 1866 'the natives continue quiet and peaceable. In no part of this colony have the early settlers been so secure from plunder or attack, not a house nor a head of cattle nor a sheep has been touched'.[4]

Severe drought, smallpox and general unrest was to help change this situation, culminating in 1868 with the murder of a police constable and some pearlers. The supposed 'manhunt which resulted' was probably instrumental in virtually decimating the Aboriginal population. In any case, there are no accounts of any Aborigines living a traditional lifestyle on the Burrup Peninsula or nearby islands after 1868.

Recent ethnographic evidence has failed to reveal any direct descendants of the Burrup Peninsula Aborigines, except one who claimed no knowledge of the area.

Although the written history reflects very little of the lifestyle of the Burrup Peninsula Aborigines, the large number of archaeological sites indicates extensive usage of the area by Aboriginal people in prehistoric times.

The coastal embayments show deflated dune systems containing habitation areas which are littered with shellfish remains common to the sandy and silty bottoms of the bays. Stone artefacts occur throughout

Deflated habitation site in Coastal Dune, Watering Cove.

these deposits, as well as European glass artefacts made from bottles which indicate that the sites were used up until recorded history. The mangrove tidal mudflat zones also have small shell dumps scattered around their fringes.

Moving away from the coast, dry stony watercourses traverse the inland plains and continue into the steep plateaux which form the Peninsula's backbone. Shells and artefacts are found scattered along these watercourse 'pathways' to large habitation sites around semi-permanent rock pools in the valleys. There are ten major valley complexes on the Peninsula and these formed the habitation focus for the Aborigines away from the coastal fringe.

Stone artefact quarries occur near rock pools and on isolated outcrops. On many flat rock surfaces there is striking evidence that seed grinding activities occurred. Many patches are abraded or smoothed through rubbing action with a grindstone. As they occur adjacent to semi-permanent pools, and some patches have glossy surfaces (possibly formed by silica polish), it seems likely that grass seeds were ground to produce large quantities of flour.

The major archaeological attraction on the Peninsula is its rock engravings, which probably number in excess of 80,000. Most appear scattered along the coast or clustered around the semi-permanent pools, where they form huge galleries.

Engraving sites close to the present coastline show a high percentage of marine oriented motifs. These include turtles, dolphins, whales, sharks and wading birds, all of which are within this environment today. Simple human figures often have exaggerated features such as feet, breasts and genitals, while some motifs suggest everyday activities such as spearing fish. A number of motifs are geometrical and non-figurative with circles, ovals, lines, arcs and curvilinear networks being present.

Engravings found away from the coast mostly occur on rock outcrops often in association with other archaeological features, such as habitation sites. There is a visible change in motif type; human and faunal motifs increase and marine-based motifs become infrequent. Huge panels of macropods are often decorated by infilling, linear bars, dots and scratches. Smaller macropods, birds, faunal tracks and highly decorated humanoid figures of many types often occur. One inland art site has in excess of 5,000 motifs clustered around freshwater pools with a medium sized habitation site nearby.

There is some evidence that Aborigines were making engravings at the time of European settlement. G. J. R. Stow, who visited Dolphin Island in 1865, states: 'the natives showed us some of their drawings on the rocks. There were sketches of fishes, turtles, lizards and different kinds of birds including emus'.[5] E. Clement, who travelled through the same area, also describes the engraving methods: 'Very rudimentary carvings are found on all hill tops and consist mainly of representations of emus, kangaroos, snakes, turtles and human beings in all sorts of positions, not a few in very vulgar positions. The carving is done with a stone axe. The design is drawn with chalk or charcoal on the rock and, by repeatedly hammering along the lines with a stone axe, it is cut deeply into the rock'.[6]

The Peninsula and the Archipelago as well must be considered as a major archaeological resource for the study of the prehistory of Australia. Human occupation of the Pilbara region suggests presence by at least 30,000 years ago, with possible earlier occupation.[7] Certainly further investigations will establish the antiquity of man in the area at a date comparable to the oldest in Australia. As noted before (Dix 1974), the archaeology of the Dampier Archipelago stands on equal footing with other notable prehistoric cultural remains found throughout the world. Its preservation must be seen as essential to Australia's cultural heritage.

(*Winter 1982*)

The Register of Significant Trees

Eve Almond and Peter Lumley

Many of the historic buildings and landscapes classifed by the National Trust owe much of their character and charm to the presence of mature trees. These trees either are remnants of original forests left to provide shade and shelter, or were planted by the early settlers in their efforts to ameliorate a hostile and unfamiliar environment. The majority of these trees are now overmature and their continued existence can no longer be taken for granted. A growing interest in the conservation of an environment previously considered alien and worthless reinforces the concern for large, old, rare, beautiful and otherwise significant trees. It is in this context that the Register of Significant Trees of Victoria has developed.

As early as 1905, N. J. Claire wrote:

'... *the giant trees now existing are few and far between, and in consequence of the little interest taken in them, either by the Government or private individuals, in the course of another half century they will have ceased to be.*'[1]

Of the 182,108 hectares of productive forest cleared for agriculture in South Gippsland, about one third was abandoned in the first 40 years and a further third was unprofitable and neglected. Where forests were harvested for timber it was often done in a wasteful and crude manner by selective logging and burning.[2] The 1939 bushfires burnt much of what remained of Victoria's virgin forests. Today the areas of unscathed bushland are greatly reduced. Among the 151 plant species reported by Leigh, Briggs and Hartley as rare, endangered or vulnerable are 10 species of eucalypts.[3]

The early settlers introduced a huge variety of exotic plants of both economic and ornamental value. Acclimatization societies were formed to test the new introductions. By the 1860s several commercial nurseries were flourishing in Melbourne and the provincial towns. T. Lang, who ran a Ballarat nursery, recorded in an extant notebook the plants he received from sources in Australia and overseas. The list is impressive and contains many species which obviously did not thrive well enough to reach the nursery catalogues.

Another important source of plants was the Botanic Gardens, Melbourne, whose Director, Dr. F. Mueller, exchanged seed with numerous overseas botanists and gardens. His report of 1869 describes the provision of 355,218 plants, mostly trees, to public reserves, cemeteries, church and school grounds throughout the colony in the years 1859–1867.[4] Characteristic species today mark his efforts in the country towns of Victoria and provide a fruitful source of nominations for the Register.

In 1980, in response to increased public concern, the National Trust of Australia (Victoria) commissioned the Royal Botanic Gardens, Melbourne, to develop a programme to identify and protect the State's notable trees. Thus the Register of Significant Trees of Victoria came into being.

Although several State bodies maintain internal lists of notable trees for their own use, the concept of a comprehensive scheme to record and

The Register of Significant Trees

Left The only known occurrence of Buffalo Sallee (*Eucalyptus mitchelliana*) is on the Buffalo granite plateau, north-eastern Victoria. The stand has been nominated because of its rarity and localised distribution. (Photo: L. Costermans)

Above The 'Separation Tree' (*Eucalyptus camaldulensis*) in the Melbourne Botanic Gardens is linked with the public celebrations in 1850 on the notification of the separation of Victoria from the Colony of New South Wales. The tree has been Classified for its historical association and as remnant vegetation in an historical garden. (Photo: E. Churcher) conserve trees of national or State importance is new to Australia. Overseas there are several comparable programmes. Since 1940, the American Forestry Association has kept the 'National Register of Big Trees' which consists of the largest reported specimens of native and naturalised trees in the United States of America. These trees are referred to as 'National Champions' and are ranked according to a formula based on height, circumference and canopy spread. Their protection depends on the public interest generated by the project.[5]

The Royal New Zealand Institute of Horticulture maintains the 'National Register of Notable and Historic Trees' which, as the name implies, has a wider charter than the American programme. Trees are registered under the following categories: objects of beauty, recognised landmarks, scientific importance, sources of rare propagating stock and historical importance. Once a tree is registered, protection may be afforded under the Town and Country Planning Bill. The Register was established in 1976 and has received support from most County and Borough Councils and several local registers have been initiated.[6]

The picture in the United Kingdom is rather diffuse with private individuals and various government agencies holding informal lists. The Department of the Environment can issue Tree Preservation Orders at its discretion, but there is no national inventory setting standards of significance.

Victoria has now taken the lead in this country with a register jointly maintained by the National Trust of Australia (Victoria) and the Royal Botanic Gardens, Melbourne.

The primary aim of the Register is to systematically record and protect outstanding trees (native or exotic, wild or cultivated) throughout the State. Trees, avenues or stands can be nominated in one or more of the following categories of significance.

1. Any tree which is of horticultural or genetic value and could be an important source of propagating stock, including specimens that are particularly resistant to disease or exposure.

2. Any tree which occurs in a unique location or context and so contributes to the landscape, including remnant native vegetation, important landmarks and trees which form part of an historic garden, park or town.

3. Any tree of a species or variety that is rare or of very localized distribution.

4. Any tree that is particularly old or venerable.

5. Any tree outstanding for its large height, trunk, circumference or canopy spread.

6. Any tree of outstanding aesthetic significance.

7. Any tree which exhibits a curious growth form or physical feature such as abnormal outgrowths, natural fusion of branches, severe lightning damage and unusually pruned forms.

8. Any tree commemorating a particular occasion or having associations with an important historical event.

9. Any tree associated with Aboriginal activities.

Following the Victorian National Trust's current practice in assessing

Above This 'Canoe Tree', Grey Box (*Eucalyptus microcarpa*), near Bacchus Marsh, has been submitted to the Register through its association with Aboriginal activities. (Photo: D. Myers)

landscapes, a two-tiered system of grading trees is used. The Register's 'Classified List' contains all those trees, avenues and stands which are of significance on a national or State basis and must be protected and conserved. The 'Recorded List' contains those trees, avenues or stands which are of interest and whose conservation and protection is encouraged. Trees of local significance are included here.

Data is stored on the computer at the Royal Botanic Gardens. This allows the handling of a large volume of information, its collation, categorization and rapid retrieval. For instance, the programme will produce a list of all Classified and/or Recorded trees from a certain Shire or a list of all trees registered under a specific category, say Category 9: 'Trees associated with Aboriginal activities'.

Once received, a nomination is held on file pending assessment: its submission is acknowledged and the nominator is asked to keep the Steering Committee which manages the Register informed of any changes in the tree's circumstance. It is envisaged that the Project Officer or other qualified person will eventually visit each nominated tree for verification and photographing. As this will be done on a regional basis, there may be some delay before a particular tree is formally rated. After a tree is checked a comprehensive brief is prepared and submitted to the Committee for appraisal.

The Committee has so far registered only a small number of trees. These have been selected as representatives of particular categories, to establish precedents and to set standards. One of the first to be Classifed was 'The Separation Tree' in the Royal Botanic Gardens. The tree, one of two remnant River Red Gums (*Eucalyptus camaldulensis*) in the Gardens, is well over 300 years old. In 1850, Melbourne citizens gathered under its widespreading branches in the newly established Gardens to celebrate the separation of the Colony of Victoria from that of New South Wales.

The single Graceful Pine (*Araucaria rulei*) in the Botanic Gardens has also been Classified. Collected on an expedition to New Caledonia in the early 1860s, this tree is believed to be the only mature specimen of its kind in Australia and hence is registered under Category 3.

Also Classified under Category 3 are the only two natural stands known of Silver Gum (*Eucalyptus crenulata*). Although the first stand is a substantial size, containing some 1,000 trees, the second area contains only 25 trees. The species is endemic to Victoria and in view of its restricted distribution it was considered appropriate to Classify both stands as a single entry.

The 'Burke and Wills Tree', a huge Moreton Bay Fig (*Ficus macrophylla*) at Swan Hill is Classified under Categories 2 and 5. The tree was planted by a Dr. Gummow, an ardent natural historian, in 1862, in his front garden. Some years later Burke and Wills stayed with the Gummows en route to central Australia. From there the association of ideas grew.

Another huge old tree which has been Classified under Category 5 is a Monterey Pine (*Pinus radiata*) at Blackwood near Bacchus Marsh. The giant stands some 49 m in height with a circumference of 8.2 m at 1.4 m above the ground.

The lovely Avenue of Honour in Bacchus Marsh itself has been Classified under Categories 6 and 8. With the construction of a by-pass around the town this avenue of Dutch Elms (*Ulmus × hollandica*) has been well preserved and forms an inviting entrance to the town.

Of local importance is the old Pepper Tree (*Schinus molle*) at the Bacchus Marsh railway station which was planted to commemorate the station's opening. The tree is one of the largest of its type in the region and is therefore Recorded but there are larger trees elsewhere so it is not of State significance.

Although the briefs on each tree are comprehensive they do not provide a 'score' for individual nominations. Committee decisions are based on the experience of its members and are to a large degree arbitrary. In the future, some form of scoring system, although still arbitrary, may help to rank borderline nominations. The ranking of historic gardens in this way by Watts has proved useful.[7]

It is certainly the intention of the Committee to set minimum standards for registration. Present indications are that the Register will contain some 500 Classified trees or stands and several times this number of Recorded trees.

When a tree is registered the owners and relevant authorities are informed and its protection urged. It is recognised that the care of trees, particularly older ones, can be costly. Unfortunately the Register cannot financially assist the owners of such trees, but where practical, expert maintenance advice is freely given. The renovation and preservation of old buildings has, for many years, attracted much attention and funding. The restoration of gardens and the conservation of trees are still relatively new concepts and funds are less easy to obtain.

As yet there is no specific legislation enacted to conserve significant trees. However, in matters of regional planning, the Third Schedule of the Town and Country Planning Act which relates to conservation and enhancement of buildings and landscapes is appropriate. At present it is considered that the most effective way of protecting trees is to make people aware of their great value so that the community itself takes the responsibility for their preservation.

From the beginning the emphasis has indeed been on community support and involvement. The general public, local government, State and semi-government bodies have been asked to nominate trees for assessment. A standardized data form has been developed and distributed throughout the State.

Since the official launching in June 1981, more than 400 submissions have been received from private individuals. Nominations from Government agencies such as the Forests Commission and National Parks are still to come. Some of the submissions were the result of the highly successful photographic competition 'Trees in Victoria' run jointly with the Ministry for Conservation during the second half of 1981 to publicise the Register.

The Steering Committee, under the auspices of the National Trust of Australia (Victoria), also reflects the community basis. As well as officers from the Trust and Gardens, it includes representatives from the Forests Commission, Garden State Committee, Ministry for Conservation, Royal Australian Institute of Parks and Recreation and the Royal Horticultural Society. Expert advisers, such as the State Historian, are consulted from time to time. Funding too has come from a variety of sources. The initial pilot study was financed by the Australian Heritage Commission. Since then a number of private companies and trusts including Australian Paper Manufacturers Ltd. and the Myer Foundation have made substantial bequests. Neil Douglas, the Australian landscape artist, has painted the two remnant River Red Gums in the Botanic Gardens, which are Classified on the Register. The painting is being used in a high quality poster; proceeds from its sales will aid the project.

The Register of Significant Trees of Victoria is still in its infancy. With 1982 being declared the Year of the Tree by the United Nations Association of Australia, it is hoped that substantial progress will be made in collating, verifying and assessing data during this year.

It is also hoped that the Register will foster public interest in trees in general, not just in Victoria but also in other States. The concept of a Register of Significant Trees does not properly belong to one region but should reflect a national concern for the identification and conservation of this common aspect of our heritage. (*Winter 1982*)

Historic public gardens, Perth

Oline Richards, A.A.I.L.A., Landscape Architect

The historic public gardens in Perth have their origins in the nineteenth century. Most of the gardens were laid out and planted during the second half of the century; the ideas behind their formation are those of nineteenth century Europe and America.

The gardens can be seen in part as an overall response to the environment, and at the same time as a reflection of various perceptions and ideal images which have been imposed on the Western Australian scene.

The gardens form part of a popular culture; many characteristics from the Victorian era have remained popular in the public gardens to the present day. Thus a garden which was planted in the city in the late 1890s and recently reconstructed as an entirely new garden using only the mature trees from the original garden, is nevertheless an historic garden and a contemporary garden rather than a Victorian revival.*

A study of the history of the public gardens illustrates the ambiguities which exist in a society and which exist in the notion of building gardens in a perverse natural environment. The fundamental ethos of a free enterprise, individualistic society, coupled with the aspirations for a comfortable lifestyle in a benign environment, on which the original settlement of Western Australia was founded have remained unchanged in many respects to the present day. The ideas underlying the gardens can be understood in an historical context and at the same time the gardens can be seen as a reflection of a contemporary society.

At its inception, the colony of Western Australia was distinctive in two respects; first it was a capitalist and elitist society; second it was the climactic expression in Australia of the pursuit of the antipodean paradise dream.[2] The colony was established solely for private investors, with only minimal economic support from the government. This was the first experiment along these lines in Australia. There were fundamental weaknesses in the concept with the result that the colony encountered serious difficulties from the outset, followed by a long period of slow growth in which there was limited prosperity.

Land was offered on generous terms; this factor combined with a belief in the superiority of the environment were a powerful attraction to settlement. That the colony's failure to prosper can be attributed to a lack of money capital and to the peculiarities of the environment is one of the many contradictions which are characteristic of the history of development in the State.[3]

In the light of the factors leading to settlement it has been suggested that the Swan River Colony is the one Australian example of a 'colony of desire', in which the pursuit was of material well-being rather than utopian ideals.[4] This pursuit of a desired environment can be seen as a persistent theme in the subsequent history of the State. A tradition, which was likewise enduring, of ungenerous public spending also manifested itself during the early phase of settlement.

At a time when other Australian cities were experiencing progress

---

* Harold Boas Gardens formerly Delhi Square.

Perth was growing slowly as a small, if pleasant, backwater.

When strolling about the streets of Perth for the first time, the stranger will notice a certain unconnected look about the different houses and Government offices. Most of the buildings are handsome and well arranged; but each one seems to stand alone, and the next neighbour to a large and well stocked 'store', or the private house of an important official, may be the cottage of a shoemaker or the yard of a blacksmith. Moreover, since all the houses in the best parts of the town stand in their own gardens, no actual streets can be said to be formed by them, and the general appearance of the whole place is rather that of those suburbs to which the business men of our large towns at home retire after their day's toil is over, than that of the hive itself. Although this impression given by the first view of Perth is doubtless disappointing to anyone who arrives with the hope of making money therein, it makes the place much prettier than it would probably be if a larger trade were carried on there. There is a look of cheerfulness and brightness about the many gardens which surround the houses and the avenue of trees which lines each side of the main road passing from one end to the other of the town, that makes the newcomer feel that a home there might be a very pleasant one.[5]

It is evident from this description of 1863 that the basic spatial framework of the future city had already been established. In addition, the quality of the loosely structured, open approach to development in the sub-tropical conditions of the local environment was to have an important bearing on the development of the city and suburbs in the late 1890s and 1900s.

Although the colony was established in 1829, it was not until the second half of the century that the public gardens began to play a part in the social and cultural life of the town. By the 1880s the capital had reached a stage of development when the problems typical of nineteenth century cities were beginning to emerge. On the eve of the gold discoveries which were to dramatically change the city there was a feeling that the earlier planning was inadequate; squares and parks were one of the recommendations for improving the quality of city life.[6]

With the advent of self government in 1890, and the development of the goldfields the State experienced a period of extraordinary growth. The capital developed rapidly during the 1890s and 1900s. It was during this time that the majority of the public gardens which exist in the city today, came into being. By the early 1890s there were two government reserves within the city boundaries, worthy of being called gardens; two decades later there were at least nineteen reserves[7] close to the city centre, being used for public parklands, gardens and playing fields.

This rapid development of the city prompted a concern for an appropriate city image. The conservatism of the early years was modified by the newly acquired prosperity and the influx of immigrants. A more radical mood entered into political life and the aspirations for the city became more ambitious. The well known parks and public gardens of Melbourne, Sydney and Ballarat[8] for example, were looked to as models of what should be aimed for in Perth in order that the city could join the ranks of Australia's capital cities.[9]

A long-term programme was embarked on for the beautification of the streets, the development of parks and gardens in the city and along the river foreshores, and the acquisition of land for childen's playgrounds and parklands.

The ideas underlying the schemes owned much to the City Beautiful and Garden City Movements in North America and England. They embraced the notion of providing 'adequate breathing spaces' in order to 'promote and safeguard' the health of the people, and were concerned with the 'ornamentation' and enchancement of the urban environment in the interests of developing civic pride and self-esteem in Perth people,

Right The people's park: strollers in Kings Park in the early 1900s. (Photo courtesy The W.A. Newspapers Ltd)

Bottom left Exotic vegetation in the Zoological Gardens. A garden 'that could not be imagined to have grown there unaided'. (Photo courtesy Mrs M. Manning)

Bottom right Avenue of red flowering gums (*Eucalyptus ficifolia*) in Kings Park (c. 1920s). (Photo courtesy The W.A. Newspapers Ltd)

particularly those of the lower orders. A belief in environmental determinism was present in the thinking of the citizens who were promoting the ideas in the early decades of this century. The aims of the civic leaders in planning for the development of the metropolis were to both emulate and to surpass the achievements of the other Australian capitals and to transform Perth into the *Garden City of Australia*.[10]

Public parks and gardens are largely a product of the nineteenth century; they originated in the cities of Europe and England in response to the changing social conditions of that period.[11] In Australia similar public parks for urban recreation were often associated with the acclimatization and botanic gardens which had a predominantly practical and scientific purpose in the earliest years of the settlements.[12]

In Perth a similar pattern was evident, although on a much smaller scale. In the initial townsite plan a garden reserve for acclimatization and horticultural uses was located alongside the Government Domain in the administrative centre of the town. As a result of public initiatives this land was later developed as a public park and botanic garden. At that time it was called the Public Gardens, although today it is know as Stirling

Gardens.[13] By the early 1890s this garden and the adjoining grounds of Government House were the main official gardens in the town. Stirling Gardens has held a dominant position in the life and the fabric of the city throughout its entire history. As a botanic garden, Stirling Gardens had only a token role; botanic gardens of the size and excellence of those in Sydney, Melbourne, Adelaide and Hobart were never formed in Perth. A proposal in 1900 to dedicate the grounds of Government House to the people of Perth, was imaginative, bold and egalitarian.[14] The Government House gardens combined with Stirling Gardens would have created an impressive major garden in the main street of the city. It is an idea which is still relevant.

In the late 1870s a second garden was being formed on the outskirts of the town, granted by the Governor of the day to the City Council; its development became his particular interest. It was appropriate therefore that the garden should be named Victoria Park. As an expression of loyalty, this was a nineteenth century tradition widespread throughout the Empire; and exemplifies the place of public gardens in the popular culture of the time.[15]

The contemporary tastes in garden design which were dominant in Victorian England were reflected in these two early gardens. These ideas were disseminated widely through popular garden literature and by the movement of people throughout the Empire and in Australia. The characteristic features were also repeated in the numerous parks and squares which developed in the period of garden building in Perth around the turn of the century. Contemporary descriptions of the gardens illustrate the complementary approaches to design which existed at the same time, and were applied according to the appropriateness to the site conditions. A geometric and axial form was used for smaller gardens and squares on level ground, and where the street grid was a dominant factor. An informal approach was typical for larger parks with a sloping terrain and particularly in association with low lying ground. Many features were however common to both forms and were essentially the same aesthetic often referred to as gardenesque.

Stirling Gardens had a small botanical collection of trees and shrubs, typical of the collections found in other Australian public gardens but with a more limited range of species, due in part to the soils and climatic conditions, and also to the small government expenditure on the garden. There were flower beds, also grass plots of couch grass, which was most suited to the hot summers. Curved gravel paths edged with brick had been overlaid on an earlier geometric grid. As more money became available it was hoped to improve the quality of the decoration and to install 'a proper iron fence along the walks'. A rustic wooden seat encircled a clump of giant bamboo, the focal element at the intersection of the main axial paths. There was also a modest conservatory.[16]

In contrast to the relative urbanity of Stirling Gardens the planning of Victoria Park was conceived in romantic terms, 'the wildness and roughness' of the natural environment was to be converted into 'an ordered and neat attractiveness'.[17]

The grounds will be tastefully laid out with shrubs and ornamental trees and will be in connection with the sericultural establishment and mulberry plantation. The tea-tree swamp flowing through into the brook (Claise Brook) is to be cleared and its waters confined in a narrow channel, so as to form a pretty rivulet terminating in a pond, around which aquatic and other plants may be grown, and the waters stocked with varieties of fish, such as have been successfully acclimatized in the neighbouring colonies.[18]

The reserve was laid out with gravel paths, flower beds, summer houses and seats under trees, and rustic bridges across the brook. A caretaker lived in a pretty little cottage by the entrance gate.[19]

The general philosophy underlying the gardenesque aesthetic was that a garden was not natural, it was both an 'artificial contrivance' and a work of art and should show artistic taste in the choice and composition of the garden's components and good manners by being honest. This could be achieved 'by using geometric layouts; by placing trees in isolation from each other, so that they were obviously planted as specimens; by composing a garden of non-native species that could not be imagined to have grown there unaided'.[20]

This aesthetic philosophy was reflected in the local approach to water and plant materials. Ponds and lakes were a source of pleasure and feature in all of the most popular public gardens.* Natural lakes and swamps are a characteristic feature of the whole of the Perth metropolitan area and were originally extensive within the townsite itself. The existence of these wetlands was an impediment to development initially, but became important at a later stage in determining the location of many of the parks and children's playgrounds. The swamps were first used as market gardens and later acquired for public reserves by the City Council, largely because the land was cheap; and also because less watering was required during the summer. These swamps became stylised lakes within a garden setting, with islands; grassed sloping banks; water lilies and ornamental grasses and stocked with exotic fish and *white* swans.

Colourful displays of massed flowers and shrubs, which are typical of municipal gardens the world over, were similarly popular. Beds and borders of annuals and roses were important elements in all the public gardens, thousands of annuals were raised in the City Council nursery and planted out twice a year in Weld Square, Russell Square and St. George's Terrace Reserve in the once select residential area at the west end of the Terrace.

Rustic features, another element in the aesthetic, suited the local shoestring budgets. Fountains and grottoes were constructed of limestone; seats, bridges and garden shelters were made of twisted timbers and lattice work. Contemporary photographs leave little doubt about how rustic many of the structures were. It is not surprising that these ornaments to the gardens were considered expendable, although the limestone work at the Zoological Gardens is a rare survival of the period and the craft. Bandstands were also typical elements, often located in the centre of the squares, and replaced the earlier rustic fountains. Band concerts held on Sundays and summer evenings during the week were subsidized by the City Council and had a popular following.[21]

Somewhat surprisingly official documents, such as annual reports, provide information on the criteria for tree planting and the selection of appropriate species. This was probably an effective way in which the attitudes and perceptions of particular individuals from a select group were able to influence other members in the community. Trees and palms were planted in clumps and avenues, Pinus, Cupressus, Ficus, and Araucaria were recommended to contrast with the lighter greens of the native species. Palms gave an 'Oriental look to the scene' and 'the straight upright character' of the Norfolk Island pine was seen to 'contrast favourably with the general looseness and straggly nature of the indigenous foliage'.

Exotic and indigenous species were planted as street trees; palms (*Phoenix canariensis*), false acacia (*Robinia pseudo-acacia*), pepper (*Schinus molle*), plane (*Platanus orientalus*), white mulberry (*Morus alba*), and Cape lilac (*Melia azederach*) were popular exotics; the indigenous species included sugar gum (*Euc. cladocalyx*), red flowering gum (*Euc. ficifolia*), blue gum (*Euc. globulus*), fig (*Ficus macrophylla*), peppermint (*Agonis flexuosa*), and silky oak (*Grevillea robusta*). Avenues of Cape lilac, sugar gums and red flowering gums were particularly liked and

---

* e.g. Hyde Park, Queens Gardens, Zoological Gardens.

planted repeatedly despite the susceptibility of the species to disease and insect attack. From a long term viewpoint this popularity was unfortunate in that the city might now possess many more avenues of mature trees than it does.

The 'approved' criteria for avenue plantings were trees with a symmetrical form, spaced at regular intervals and of a uniform kind. There was in this preference a strong bias towards order and integration which was not always in conformity with the general public sentiment. Eucalypts did not generally meet this specification, but were defended forcefully on the grounds that they symbolized an Australian identity. Some private citizens also appeared to prefer variety and the opportunity to determine their own environment by planting the trees of their choice on the verges outside their houses.[22]

Gardens and parks which were representative of those developed in the 1890s and 1900s and still in use today, are Kings Park, The Zoological Gardens, Russell and Weld Squares, The Esplanade, and Perth Oval. In none of these gardens has there been a deliberate intent to conserve them for historic reasons. The condition of the gardens tends to reflect the status of the surrounding neighbourhood and are well maintained or have been allowed to decline accordingly.

Kings Park is the most notable of the historic gardens; it has always had a special significance in the social and cultural life of Perth. It was the gift of the Forrest Government to the people of Perth in 1894, a grand and democratic gesture in keeping with the spirit of the time. It is a large park, one thousand acres (400 ha) overlooking the city. It has developed over the years as a public playground and a national shrine, and contains virtually the entire collection of the city's monuments and much historic memorabilia. Since the beginning of settlement it has been a favoured spot for admiring the spectacle of the Swan River and recording the progress of the City.

Kings Park is a good illustration of the conflicts which exist in a society dominated by individualistic values. This conflict is clearly expressed in ambiguous attitudes toward public and private domains and to individual and collective interests. Proposals to locate both the University and facilities for the Empire Games in the park and which were subsequently abandoned, brought to the forefront the depth of feeling in the community on the rights of all the people to have free access to the most attractive and desirable urban open places.[23]

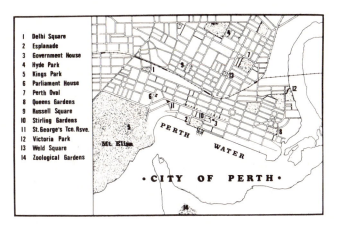

Top Public gardens and popular culture: Zoological Gardens, South Perth (c. 1900).

Above The rustic was a part of Victorian aesthetic tastes. Zoological Gardens, South Perth. (Photos courtesy Mrs M. Manning)

Despite this feeling, the appropriation of public lands for the use of particular groups is universal. Most public parks have portions set aside for private club facilities, and in some instances particular interests dominate the use of a reserve.* There also exists the symbolic appropriation of public lands by individuals, particularly the most affluent, which is demonstrated in a proprietary attitude toward public property by landowners whose properties overlook parks and scenic views.

The Garden City images that were first promoted some eighty years ago have become built into the perceptions of the city.[24] The preferences in the early settlement for an environment with a garden character, which was quite unlike the European cities of the nineteenth century, is reflected in today's urban environment, and could well have been influenced in part by the scenic attractiveness of the townsite and the nature of the Mediterranean climate.

In addition to the mosaic of open spaces within the city itself, many public buildings which have survived from the late nineteenth and early twentieth century still stand in garden settings.† While the majority of the buildings from this period have disappeared the city open spaces remain largely intact.

This raises the issue of the conservation of historic public gardens, and poses some difficulties. It is in the nature of gardens that they are constantly changing; the gardens experienced by the people of Perth in the 1900s are not the same gardens being experienced today. To understand the gardens it is necessary to attempt to place them in an historical context, and to interpret the relationships between the gardens and Perth society.

Because public gardens are a product of an urban culture, and reflect generally held attitudes and popular tastes, they are susceptible to the changes which are typical of modern urban societies. If the gardens are to survive as vital components in the urban scene they must respond to changes. At the same time an awareness of the history of the gardens and speculation on their meaning are one aspect of maintaining their value in a society. (*Winter 1982*)

---

* e.g. Dorrien Gardens.

† e.g. Wesley Church, Perth Railway Station (Wellington St. garden), St. George's Cathedral, Parliament House, Perth Public Library.

Notes and references

Wright Towards the bicentennial landscape

References
1. Bernard Smith, *European vision and the South Pacific*, Oxford University Press, London, 1960. p. 134.
2. Observations on the disappearance and probable cause, of many of our native birds in Central Queensland, *The Queensland Naturalist* IX, 1, 1934.
3. Memories of far-off days, *MS A327*, Dixson Library, University of New England.
4. L. Webb, The rape of the forests, in A. J. Marshall, ed., *The great extermination*, Heinemann, Melbourne, 1966.
5. For example, Eric Rolls, *They all ran wild*, Angus and Robertson, Sydney, 1969, pp. 353–6.
6. See for example, *The National Farmer*, Oct. 2, 1980, pp. 14–15.
7. Ibid, p. 14.
8. *They all ran wild*, pp. 58–60.
9. *Report of the National Estate*, A.G.P.S. Canberra, 1971, pp. 62–73.

Other references
David Denholm, *The colonial Australians*, Allen Lane, Ringwood, Vic., 1979, especially for the effects of staight line survey methods.
L. J. Webb, *Environmental boomerang*, Jacaranda, Brisbane, 1973.

Fox Kakadu — man and landscape

Further Reading
Blainey, G. (1966) — *The Tyranny of Distance*, Sun Books, Melbourne.
Carrington, F. (1885) — 'Report on lands at South and East Alligator Rivers', South Australian Lands Office: Correspondence Dockets. N.T. Series, 1868–1910.
Cole, K. (1975) — *A History of Oenpelli*, Nungalinya Publishers, Darwin.
(1980) — *Dick Harris: Missionary to the Aborigines*, Keith Cole, Darwin.
Herbert, X. (1975) — *Poor Fellow My Country*, Collins, Sydney.
Leichhardt, L. (1847) — *Journal of an Overland Expedition in Australia*, Facsimile edition, Doubleday, Sydney.
Macknight, C. C. (1976) — *The Voyage to Marege*, Melbourne University Press.
Meehan, B. (1980) — 'Eclecticism of wetland hunters under threat' *in* Status and Management of Tropical Wetlands Workshop. Unpublished Report, ANPWS, Canberra.
Michener, J. A. (1959) — 'From the Farm of Bitterness' in *Hawaii*, Corgi, London.
Warburton, C. (1944) — *Buffaloes*, Consolidated Press, Sydney.

(I would like to thank Nipper Kabirriki who told the Bardmardi story and to George Chaloupka who recorded it. This version is very, very much abbreviated and any serious omissions or errors are mine. A.M.F.)

Hassall Movables of heritage significance

References
1. *Customs (Prohibited Exports) Regulations* Second Schedule, Items 1, 1B, 2, 5A, 5B, 5C and 14.
2. Lockhart, J. S., *Occasional Address at the Conferring of the Degree of Bachelor of Laws*, University of Sydney, 27 February 1982.
3. See Prott, L. V., & O'Keefe, P. J., *National Legal Control of Illicit Traffic in Cultural Property*, UNESCO, Paris, 1983.
3A. See *Commonwealth* v. *Tasmania & Ors.* High Court of Australia, 1 July 1983. (Franklin Dam case)
4. e.g., *Attorney-General of the Duchy of Lancaster* v. *G. E. Overton Farms Ltd.* reported in *The Times*, 26 June 1980. (debased silver Roman coins held not to be Treasure Trove)
5. *Antiquities Act* 1982 (U.K.).
6. Waller, *Coronial Law and Practice in New South Wales*, Sydney, 1982, pp. 3, 177–8; and see Garland, H. K., and Jones, D. J., *Detecting Australian Treasure*.
7. (1684) 1 Vern. 273.
8. *Dougan* v. *Ley* (1946) 71 CLR 142, per Dixon, J.

Fahy Andrew Lenehan — Sydney cabinetmaker

Acknowledgements
The text and illustrations are drawn from *19th century Australian Furniture*, by Kevin Fahy, Christina and Andrew Simpson.
The authors wishes to acknowledge the following sources: C. Craig and others, *Early colonial furniture in New South Wales and Van Diemen's Land*, Melbourne, 1972.
K. Fahy, 'Andrew Lenehan — cabinetmaker', *Descent* vol 6, pt 1, 1972.
The Historic Houses Trust, *Sydney's colonial craftsmen*, Sydney, 1982.
together with M. Graham, W. Chapman, R. Lenehan, the Mitchell Library, the State Archives of Queensland and others.

Flood South West Tasmania

Bibliography
Angus, M. *The World of Olegas Truchanas*. O.B.M. Pty Ltd, 1975.
Australian Conservation Foundation. *The Wonderful Southwest*. Habitat, Special Issue, 1975.
Bell, C. and Sanders, N. *A Time to Care*. Chris Bell, Blackmans Bay, 1980.
Boss-Walker, Ian. *Peaks and High Places: Cradle Mountain — Lake St. Clair National Park Tasmania*, Scenery Preservation.
Dombrovskis, P. and Miller, E. *The Quiet Land*. Peter Dombrovskis Pty Ltd, Sandy Bay, Tasmania.
Fudali, R. F. and Ford, R. J. 'Darwin glass and Darwin crater', *Meteoritics*, vol. 14 (3), 1979, pp. 283–96.
Gee, H. and Fenton, J. (eds). *The Southwest Book: A Tasmanian Wilderness*. Australian Conservation Foundation, 1978.
Gowland, R. and K. *Trampled Wilderness: The History of South West Tasmania* (2nd Edition). Richmond and Son Pty Ltd, Devonport, Tasmania, 1976.
Hydro Electric Commission. *Lower Gordon River Scientific Survey* (24 reports), 1978.
Jones, R. 'The extreme climatic place?' *Hemisphere*, vol. 26 (1), 1981, pp. 54–9.
Jones, R. 'Submission to the Senate Select Committee on South West Tasmania'. *Australian Archaeology*, vol. 14, 1982, pp. 96–106.
McKelvey, M. *Cradle Country: A Tasmanian Wilderness*. Rigby, 1976.
Nielson, D. *South West Tasmania — A Land of the Wild*. Rigby, 1975.
South West Tasmania Resources Survey, Resource Inventory, Volumes 1 and 2 (Ed. P. Waterman), 1981: S.W.T.R.S..
Also: 25 Discussion Papers, 22 Working Papers, 20 Occasional Papers

Films
'Franklin River Journey'. Tasmanian Film Corporation, colour — 30 minutes.
'The Last Wild River'. Paul Smith, 1978, colour — 28 minutes.
'Climbing Frenchman's Cap'. Tasmanian State Film Unit, 1969, colour — 6 minutes.
'Tasmanian Wilderness'. Sims, 1973, colour — 102 minutes (Parts 1 and 2).
'Walk into Wilderness', Impala Films, 1973, colour — 24 minutes.
'The Franklin — Wild River'. Mike Cordell, colour — 46 minutes.
'A Wilderness in Question'. Damon Smith, colour — 46 minutes.
Tape-slide kits, Tasmanian Wilderness Society (129 Bathurst Street, Hobart, TAS. 7000).

Sherry and Baglin Australian themes in stained glass

Acknowledgements
We would like to thank the following for permission to photograph and reproduce stained glass: St. Augustine's Church of England, Hamilton, Brisbane; St. Bede's Church of England, Drummoyne, Sydney; owners of private houses (Figs. 3, 4); Yanco Agricultural High School, Yanco, N.S.W.; Cranbrook School, Bellevue Hill, Sydney; Department of Works, Brisbane; Adelaide Stock Exchange; Sydney Town Hall; Melbourne Town Hall.

References
1. For some comments on the history of stained glass in Australia, see my article, 'Secular Stained Glass in Australia,' *The Australian Antique Collector*, No. 26 (June–December 1983), pp. 44–9. Detailed research on the Melbourne firms is contained in Geoffrey M. Down's thesis, *Nineteenth-Century Stained Glass in Melbourne* (M.A. University of Melbourne, 1975) and on the Sydney firms in Danute Illuminata Giedraityte's thesis, *Stained and Painted Glass in the Sydney Area c.1830–1920* (M.A. University of Sydney, 1983). I am grateful to Miss Giedraityte for information generously offered in conversation.
2. On this revival, see A. Charles Sewter, *The Stained Glass of William Morris and his Circle* (New Haven: Yale University Press, 1974); Martin Harrison, *Victorian Stained Glass* (London: Barrie & Jenkins, 1980); and Michael Donnelly, *Glasgow Stained Glass* (Glasgow: Glasgow Museums and Art Galleries, 1981).
3. Montgomery's papers on stained glass are published in: *The Australasian Builder and Contractors' News* (ABCN), Oct. 19, 1889, pp. 368–70; *ABCN*, Sept. 8, 1894, pp. 101–03; *Building, Engineering*

and *Mining Journal (BEMJ)*, Sept. 10, 1898, pp. 271–3; *BEMJ*, Sept. 17, 1898, pp. 281–4; *Royal Victorian Institute of Architects' Journal*, vol. 5 (1907), pp. 140–57; *Year Book of Victorian Art, 1922–23*.
4. *Building and Engineering Journal*, May 9, 1891, p. 179. See also: *The Sydney Mail*, Apr. 29, 1871, p. 281; *ABCN*, Sept. 17, 1887, p. 300; *ABCN*, Apr. 20, 1889, p. 379; *ABCN*, June 29, 1889, p. 605.
5. *ABCN*, Jul. 20, 1889, p. 56.
6. *ABCN*, June 2, 1888, p. 371.
7. Their tender and accounts for this work are in the Sydney Town Hall Archives.
8. 'Mr. John L. Lyon,' *The Australasian Decorator and Painter*, Aug. 1, 1909, p. 264.
9. See Down, *Nineteenth-Century Stained Glass in Melbourne*, p. 64, and M. C. Dobson, 'John Lamb Lyon and Francis Ernest Stowe,' *Leichhardt Historical Journal*, Jul. 1973, p. 6.
10. *The Three Voyages of Captain James Cook* (London: William Smith, 1842), vol. I, p. 240. Lyon could have had access to one of the numerous editions of the *Voyages* published before 1870.
11. *The Commonwealth Home*, Feb. 1, 1929, p. 20. John Ashwin & Co., established in Dixon Street, Sydney, in 1911, was a separate company from F. Ashwin & Co. of Pitt Street.
12. Accounts and correspondence relating to this work are in the Queensland State Archives. George Gough began in business in Brisbane with R. S. Exton in 1888 but by 1896 had set up independently as George Gough & Co., advertising as 'Decorator, Sign Writer, Paper hanger, lead light worker, and artist in stained glass' (*Pugh's Almanac and Queensland Directories*, 1899, p. 653); by 1912 the company was George Gough & Son and continued to be listed in the Brisbane directories until 1936.
13. See Kenneth Goodwin, 'Morris & Co.'s Adelaide Patron,' *Art and Australia*, vol. 9 (1971), pp. 342–5.
14. British imperialism, though, seems to be a message of the window. South Australia has a wealth of stained glass but it is strongly English in orientation and relatively few Australian themes are to be found. It is characteristic, for example, that windows designed by Adelaide artist, E. F. Troy, at the time of Australia's Federation include the heraldic series in Government House, Adelaide, in honour of the visit of the Duke and Duchess of York in 1901, the window portraying Edward VII in the Adelaide Town Hall (1902), and the British Empire windows in the South Australian Institute of Technology (1903). I am grateful to Mr. Peter Donovan, who is researching stained glass in South Australia, for information which has confirmed this view.
15. 'Centennial Hall Windows,' *ABCN*, Dec. 21, 1889, p. 604.
16. Ibid., pp. 603–04. On Henry's career, see Margaret Bettridge, *Australian Flora in Art from the Museum of Applied Arts and Sciences, Sydney* (South Melbourne: Sun Books, 1979), pp. 9–11.

Kerr Cemeteries — their value and conservation

Further Reading
On Australian attitudes to death: Griffin & Tobin, *In the Midst of Life*, Melbourne University Press, 1982.
On cemetery monuments (well illustrated): Gilbert, *A Grave Look at History*, Ferguson, Sydney, 1980.
On the development of cemeteries and monuments in N.S.W.: *In Memoriam*. Published by the Historic Houses Trust of N.S.W. in conjunction with the exhibition *The Victorian Way of Death*, Sydney, 1981.
On planning work in existing cemeteries: *The Conservation Plan*, National Trust of Australia (N.S.W.), Sydney, 1982. *A guide to the Conservation of Cemeteries*, National Trust of Australia (N.S.W.), Sydney, 1982.
On the principles of conservation: *The ICOMOS Australia Charter for the Conservation of Places of Cultural Significance* (Burra Charter).

Fox Sydney's first skyscrapers

References
1. Irvin, Eric, *Theatre comes to Australia*, University of Queensland Press, St. Lucia 1971.
2. Ibid., p. 50.
3. Levi, J. S. and Bergman, G. F. J., *Australian Genesis*, Rigby, 1974.

Spearritt Twentieth century buildings

Further reading
Many of the arguments for preservation advanced here, including the criteria for preservation, are expounded at greater length in D. N. Jeans and P. Spearritt, *The Open Air Museum: The Cultural Landscape of NSW* (George Allen and Unwin, 1980). A brief history of preservation in Australia is Max Bourke's 'The Preservation of Historical Places', *Victorian Historical Journal*, vol. 51, no. 1, February 1980. Australia's rich heritage of vernacular architecture has yet to attract much attention. Meanwhile, R. J. S. Gutman and E. Kaufman's *American Diner* (Harper and Row, 1979) should whet the appetite.

Freeman Mulwala homestead complex

References
1. Thomas Townsend, 'Map of the Murrumbidgee Squatting District, 1850', Archives Office of N.S.W.; Billis, R. V. and Kenyon, A. S., *Pastoral Pioneers of Port Phillip*, Melbourne, 1932, p. 306; *New South Wales Government Gazette*, 30 September 1848.
2. John Alexander Sloane, 'A Changing World: Mulwala and Savernake Stations, 1862 to 1953', MS in possession of Mr. W. H. Sloane, Melbourne, p. 6.
3. Mrs. Gibson's Journal, quoted in Sloane, 'A Changing World . . .', pp. 15, 16.
4. Ibid., p. 10.
5. *Town and Country Journal*, 21 May 1887.
6. Sloane, 'A Changing World . . .', p. 20.
7. Ibid., p. 26.
8. 'A Tour to the South: Corowa and Wahgunyah', *Town and Country Journal*, 18 May 1872, p. 625.
9. Sloane, 'A Changing World . . .', p. 10A.

Lucas Harrisford, Parramatta

References
1. *The King's School Old Boys' Union Magazines, 1980*. Harrisford, A Brief History, by L. D. S. Waddy.
2. Ibid.
3. Ibid.
4. *Harrisford*. A coloured brochure prepared by The King's School Old Boys' Union, 1982.
5. *N.S.W. Calendar and Directory 1832*.
6. L. D. S. Waddy, op. cit.
7. *The King's School 1831–1981* by Lloyd Waddy, Council of the King's School, 1981, p. 27.
8. *The King's School Parramatta, Register 1831–1981*. Council of the King's School, 1982.
9. Information from A. R. Naylor, bricklayer during restoration 1980.
10. Report on wall samples from 'Harrisford' Parramatta by Dr. G. S. Gibbons. N.S.W. Institute of Technology, August, 1981.
11. Lloyd Waddy, op. cit.
12. Harrisford coloured brochure, op. cit. *N.S.W. Calendar and Directory*, op. cit.
13. *Australian Dictionary of Biography*, Vol. 6 — Woolls.
14. Harrisford coloured brochure, op. cit.
15. *Interview with Miss Margaret Griffiths*, Cumberland Argus 1933.
16. Conversation with Mrs. Warren Peel nee Harris, Parramatta 1981.
17. The tank had been filled in *c*. 1922 by Henry Harris, op. cit. conversation with his daughter, Mrs. Peel.

Lucas Lyndhurst – a battle won

References
1. Account of Henry L. Cooper. *Macarthur Papers*, Mitchell Library, Sydney.
2. Letter from Emily Macarthur to her aunt, p. 46, Rachel Roxburgh, *Early colonial houses of New South Wales*. Ure Smith/National Trust of Australia (N.S.W.), Sydney, 1974.
3. James Broadbent, 'Early Sydney houses, examples of pattern-book architecture', *Art Association of Australia: Architectural Papers*, 1976.

Kerr Edmund Blacket's church architecture

References
1. State Archives of New South Wales: 'Colonial Secretary, Letters Received, 1843 — Bishop of Australia', bundle 4/2589.2.
2. E. T. Blacket to Frank Blacket, 22 December, 1849. ML Doc 697.
3. B. F. L. Clarke, *Anglican Cathedrals outside the British Isles*, London, S.P.C.K., 1958, p. 89: quoted in Joan Kerr and James Broadbent, *Gothick Taste in the Colony of New South Wales*, Sydney, David Ell Press, 1980, p. 67.
4. Bishop Broughton's Report to the Diocesan Societies of the S.P.G. and S.P.C.K., 19 March, 1849: published in the *Sydney Guardian*, vol. 1, no. 11 (April 1849), p. 73.
5. Still extant in the University of Sydney archives.
6. E. T. Blacket to Frank Blacket, 12 May, 1843: ML Doc 697.
7. Ibid.
8. W. H. Walsh, 'The Ecclesiology of New South Wales': *The Ecclesiologist*, London, vol. xii, no. lxxxii (April 1851), p. 261.
9. W. H. Walsh to Rev. A. M. Campbell, Secretary, Society for the Propagation of the Gospel, 6 October, 1840: U.S.P.G., London, C.M.S. (ML FM4/56a).
10. Letter to Editor, *Sydney Morning Herald*: quoted in L. M. Allen, *A History*

269

of Christ Church S. Laurence, Sydney, Sydney, 1940, p. 10.
11. E. T. Blacket to Frank Blacket, 12 May, 1843; ML Doc 697.
12. W. H. Walsh, 'The Ecclesiology of New South Wales' op. cit., p. 263. For Christ Church choir see also *Ecclesiologist*, vol. v., no. x (April 1846), p. 158.
13. J. Horbury Hunt to Editor, *Sydney Morning Herald*, 1 December, 1889: from newspaper cutting book kept by Hunt now in the Mitchell Library: quoted J. M. Freeland, *Architect Extraordinary: The Life and Work of John Horbury Hunt: 1838–1904*, Cassell Australia, 1970, p. 27.
14. W. G. Broughton, *Visitation Journal 1845*, London, 1847: entry for 10 September, 1845, p. 32.
15. E. T. Blacket, 'Diary': entries from 24 March to 13 May, 1843: University of Sydney Archives.
16. E. T. Blacket to Frank Blacket, 12 May, 1843: ML Doc 697.
17. See *Southern Queen*, Sydney, 3 May, 1845, p. 267, which calls the maker 'Walls of Newcastle-upon-Tyne'.
18. (Richard Thompson), 'Progress of the Oxford Heresy — St. Laurence's Church': *Commercial Journal*, Sydney, 27 September, 1845.
19. The return of the window to England was reported in the *Sydney Morning Herald*, 16 September, 1845. When it came back to Sydney Blacket went to inspect it 'with the Bishop and Mr. Metcalfe'. (See 'Diary': entry for 14 July, 1852 — Sydney University Archives).
20. Hardman Papers, City of Birmingham Archives: Indexes — First Glass Day Book 1845–53, and Letters 1854, Butterfield to Hardman, 31 October, 1853. Also E. T. Blacket, 'Diary': entry for 8 October, 1852, Sydney University Archives.
21. Inscription in two rebound sets of *Architectural Ornaments*, ML D217: 'Presented to Edmund Blacket Esq., a token of regard from his very *firm* friend Dr. C. Nicholson 1856'.
22. Information from D. Giedraityte, who is preparing an M.A. thesis on nineteenth-century glass in Sydney at the Power Institute of Fine Arts, University of Sydney.
23. Cyril Blacket, 'Church Architecture. St. John's Glebe': *Sydney Morning Herald*, 15 May, 1926: ML newspaper cuttings book, vol. 233. The Australian glass in Goulburn Cathedral is by Falconer and Ashwin and Lyon and Cottier, both Sydney firms, while the English glass is by Hardman Brothers of Birmingham. For their respective contributions see Ransome T. Wyatt, *St. Saviour's Cathedral Goulburn — Jubilee 1884–1934*, Goulburn, 1934, pp. 11–14.
24. E. T. Blacket to Frank Blacket, 12 May, 1843: ML Doc 697.
25. J. M. Freeland in *Architect Extraordinary*, op. cit., p. 103, states that Hunters Hill is the only church known to have been designed by Hunt for the Anglican diocese of Sydney, although he attributes St. Augustine's at Neutral Bay to Hunt's hand on p. 105. Both were late works, the former dating from 1885 and the latter from 1887.
26. Canon Taylor was so appalled at the High Anglican acceptance of Transubstantiation, or the 'Real Presence' of Christ in the Eucharist, that he had carved on his altar the words: 'He is Risen. He is not here'. The church subsequently achieved some notoriety as 'The Church of the Real Absence'. This altar was designed by Cyril Blacket in 1895.
27. 'Architects and Architects': *Australian Churchman*, Sydney, 22 December, 1881, p. 138.
28. Probate was granted on Blacket's estate on 9 May, 1883 and was sworn in at £3,100. (Information from the *Australian Dictionary of Biography*, Research School of Social Sciences, Australian National University.)
29. 'A Clerk of Works' (subsequently identified as Turner), *Church of England Guardian*, Sydney, 19 November, 1888.
30. Ibid., 22 December, 1888, p. 6. Turner referred only to the churches; he avoided naming their architect.
31. 'A Clerk of Works', *Church of England Guardian*, Sydney, 19 November, 1888.
32. *Macleay Chronicle*, 24 March, 1881: from J. Horbury Hunt newspaper cutting book. ML Q620.8, p. 38.

Blackmore Carpenter's decoration

References
1. Fowles, Joseph, *Sydney in 1848*, (orig. 1848) facsimile, Ure Smith, Sydney, 1973.
2. Demography Bulletin No. 67, 1949, pp. 154–5, Commonwealth Bureau of Census and Statistics, Canberra, 1949: in M. Clark, *Select Documents in Australian History*, Vol. II, Angus and Robertson, 1977.
3. Goodlet and Smith Sydney catalogue of 1890.

Proudfoot First Government House, Sydney

References
1. This copper plate was found in 1899 by a workman digging a trench for a telephone cable along the Bridge Street footpath. It is now in the Mitchell Library. It has a beautiful greeny-grey patina.
2. John White, *Journal of a Voyage to New South Wales*, 1790, p. 119. Phillip's despatch 4 March 1790, *Historical Records of New South Wales*, I, ii.
3. *Historical Records of Australia*, I, 1, p. 43.
4. Captain Watkin Tench, *A Complete Account of the Settlement at Port Jackson in new South Wales*, London, 1793, p. 11.
5. Ibid., p. 10.
6. There has been pipe-clay found in the mortar on the site, but no trace of lime in the mortar of Phillip's house, when tested by Dr. George Gibbons recently.
7. Public Record Office Papers, CO 201/44(1) p. 38, microfilm in Mitchell Library. This plan was also reproduced in *Historical Records of New South Wales*, Vol. 6, p. 765. To see a reproduction of a 'Floor Cloth' of the type used at this time, made of canvas treated with a glossy finish, one should visit Vaucluse House Sydney, which has one recently specially made.
8. Mortimer Lewis's Report of 15 September 1845 for a Board of Survey of the state of the building. Archives office of N.S.W., Colonial Audited Papers 4/2717.2.
9. Letter from John Harris to Mrs. King, King Papers, Vol. 8, pp. 244–5, Mitchell Library.

Egloff Port Arthur Historic Site

References
1. M. Weidenhofer, *Port Arthur: A place of Misery*, Oxford University Press, 1981, p. 125.
2. *Report to the Inspector of Public Buildings* by H. Bucirde, Clerk of Works, 2.7.13. The Archives Office of Tasmania, P.W.D. 51/40, 042/1348.
3. Interview of Eddie Bellette, recorded 24.4.81 by Richard Morrison, Port Arthur Conservation Project.
4. J. Allen, 'Port Arthur Site Museum, Australia: its preservation and historical perspective', *Museum*, vol. 28, no. 2, pp. 99–105.
5. A. G. L. Shaw, *Convicts and the Colonies*, Faber and Faber, 1966, p. 212.
6. M. Weidenhofer, op. cit., p. 96.
7. A. G. L. Shaw, op. cit., p. 285.
8. I. Brand, *Penal Peninsula*, Jason Publications, 1978, pp. 43–4, 65.
9. R. I. Jack, 'Tasman Peninsula'. *The Heritage of Australia*, pp. 7/70 – 7/74, The Macmillan Company of Australia, 1981.
10. A. G. L. Shaw, op. cit., p. 353.
11. P. Bolger, *Hobart Town*, A.N.U. Press, 1973.
12. Crawford de Bavay and Cripps with Fowler, England and Newton, *To Conserve Port Arthur*, Report on the Conservation of Building Fabric at Port Arthur for the National Parks and Wildlife Service, Tasmania, 1979 (unpublished).
13. A. Lister, *Garden Point 1978*, for National Parks and Wildlife Service, 1978 (unpublished).

Wood Along the Ghan track

Acknowledgements
Much of the historic material comes from the following sources:
Ian Mudie, *The heroic journey of John McDouall Stuart*, Angus and Robertson, Sydney, 1968.
Basil Fuller, *The Ghan: the story of the Alice Springs Railway*, Rigby, Adelaide, 1975.
Baldwin Spencer and F. V. Gillen, *Across Australia: Vol. 1*, Macmillan and Co., London, 1912.
T. G. H. Strehlow, *Journey to Horseshoe Bend*, Angus and Robertson, Sydney, 1969.
Roma Dulhunty, *Where the dead heart beats Lake Eyre lives*, Lowden Publishing Co., Kilmore, Vic., 1979.
Doris Blackwell and Douglas Lockwood, *Alice on the line*, Rigby, Adelaide, 1965.
Howard Pearce, *Remote sites documentation: Report 1*, S.A. Aboriginal and Historic Relics Administration, Adelaide, 1980.
South Australian Archives.
Particular thanks to Peter Forrest for his efforts on the trip.

Marquis-Kyle Palma Rosa, Brisbane

References
1. Carolyn Cox, *Andrea Stombuco*, Bachelor Thesis, University of N.S.W. School of Architecture, 1977.
2. *Brisbane Courier*, 16 December 1887.
3. F. E. Lord, 'Brisbane's Historic Homes XC1: Palma Rosa', *The Queenslander*, 28 January 1932, p. 28.
4. Ray Sumner, *More Historic Homes of Brisbane*, Brisbane: National Trust of Queensland, 1982, p. 63.
5. *Brisbane Courier*, 16 December 1887.

O'Connor
Historic bridges of Australia

References
1. Birmingham, J., Jack, I., Jeans, D., *Australian Pioneer Technology*, Heinemann, 1979, p. 200.
2. Fraser, D., *Two Whipple Trusses in New South Wales*, Trans. I.E. Aust., Vol. CE23, No. 4, November 1981, pp. 272–82.
3. Jensen, E. and R., *Colonial Architecture in South Australia*, Rigby, 1980, p. 888.
4. O'Grady, V., *the Development of the Bridge in Australian Road Construction*, J. Roy. Aust. Hist. Soc., Vol. 66, Part 4, March 1981, pp. 273–86.
5. Smith, R., *Early Tasmanian Bridges*, Foot and Playstead, Launceston, 1969, p. 107.
6. *The Roadmakers*, Dept. of Main Roads, N.S.W., 1976, p. 335.

Flood The rock paintings of Cape York

Further reading
Flood, J. M., 'Quinkan Country: the archaeology and art of Cape York Peninsula', *Hemisphere*, vol. 27 (5), 1983.
Rosenfeld, A., Horton, D. and Winter, J., *Early Man in North Queensland*, Terra Australis 6, Australian National University, Canberra, 1981.
Trezise, P. J., *Quinkan Country*, Reed, Sydney, 1969.
Trezise, P. J., *Rock art of South-East Cape York*, Aust. Institute of Aboriginal Studies, Canberra, 1971.
Trezise, P. J. and Roughsey, D., *The Quinkins*, Collins, Sydney, 1978.
Trezise, P. J., *Turramulli the Giant Quinkin*, Collins, Sydney, 1982 (and other children's books). (Quinkin is an alternative spelling of Quinkan, and is now preferred by Trezise as being closer to the sound of the Aboriginal word.)

Whitmore South East Queensland coal mines

References
1. J. Gregson, *The Australian Agricultural Company, 1824–1875*, Angus and Robertson, Sydney, 1907.
2. M. H. Ellis, *A Saga of Coal*, Angus and Robertson, Sydney, 1969.
3. R. L. Whitmore, *Coal in Queensland: The First Fifty Years*, University of Queensland Press, Brisbane, 1981.

Green Archaeology of the Burrup Peninsula

References
1. J. Masefield, *Dampier's voyages*, 1906, p. 435.
2. P. P. King, *Narrative of a survey of the intertropical and western coasts of Australia 1818–22*, Library Board of South Australia, Adelaide, 1969, p. 38 (facs. ed.).
3. F. T. and A. C. Gregory, Mr. Padbury's expedition to Nicol Bay, *Exploration Diaries* v. 54, 1969. *Journals of Australian Exploration*, Library Board of South Australia, Adelaide. (Facs. ed. 1884) pp. 55–97.
4. R. J. Sholl, *Inquirer*, Perth, 21 March, 1866.
5. F. Stow, *Voyage of the 'Forlorn Hope'*, Sullivans Cove Press, Sydney, 1981, p. 66.
6. E. Clement, Ethnographic notes on the Western Australian Aborigines, *Internationales Archiv Fur Ethnographie* XVI, sheet 1 and 2, 1903, p. 9.
7. L. Maynard, A Pleistocene date from an occupation deposit in the Pilbara Region, Western Australia, *Australian Archaeology*, 1979.

Almond and Lumley
The Register of Significant Trees

References
1. Claire, N. J., 'Notes on the Giant Trees of Victoria', *The Victorian Naturalist 21*, pp. 122–8, 1905.
2. Webb, L. J., Whitelock, D. and Brereton, J. Le G., eds, *The Last of Lands*, Jacaranda Press, Milton, Queensland, 1969.
3. Leigh, J., Briggs, J. and Hartley, W., *Rare or Threatened Australian Plants*, Australian National Parks and Wildlife Service Special Publications 7, 1981.
4. *Report of the Government Botanist and Director of the Botanic Gardens 1868–1869*, Government Printer, Melbourne, 1869.
5. Behlen, D. McK., 'Big Trees: Something to Look For', *Journal of Arboriculture 6*(2), pp. 44–8, 1980.
6. Raethel, D. and Aldous, D. E., 'Notable and Historic Tree Registration in New Zealand — A Postal Survey', *Annual Journal Royal New Zealand Institute of Horticulture* 7, 54–9, 1979.
7. Watts, P., 'Finding and Assessing Historic Gardens', *Proceedings of the First Garden History Conference*, National Trust of Australia (Victoria), Australian Heritage Commission, pp. 39–42, Melbourne, 1980.

Richards
Historic public gardens, Perth

References
1. The full quotation is 'Citizens of Perth, follow me and I will make this city a fairer Athens and a freer Rome'. From a Speech made by W. G. Brookman, Mayor of Perth 1900–01, and quoted in C. T. Stannage, *The People of Perth*, Perth City Council, 1979.
2. J. M. R. Cameron, *The Colonization of Pre-Convict Western Australia*, Thesis, Doctorate of Philosophy. University of Western Australia Geography Department, 1975.
3. Pamela Statham, *The Tanner Letters: A Pioneer Saga of Swan River and Tasmania 1831–1845*. University of Western Australia Press, Nedlands, 1981.
4. Cameron, *The Colonization of Pre-Convict Western Australia*.
5. Mrs. Edward Millet, *An Australian Parsonage or The Settler and The Savage in Western Australia*, Edward Stanford, London, 1872.
6. *West Australian*, 23 Nov. 1885, leader.
7. City of Perth, *Municipal Handbook 1913*, listed under Park and Reserves: Queens Gardens, Delhi Square, Russell Square, Weld Square, Perth Oval, Forrest Park, Royal Park, St. George's Terrace Reserve, Esplanade, Wellington Square, Mulberry Plantation (Haig Park), Victoria Park, Hyde Park, Beaufort Reserve, Wittenoom St. Reserve; also there were Kings Park, Stirling Gardens, The Zoological Gardens and Monger's Lake which were not vested in the Perth City Council.
8. V. Hughes, *History of Kings Park*, Honours Thesis, Murdoch University, 1978. Also the parklands around Lake Wendouree in Ballarat were referred to by the Town Clerk of Perth, W. E. Bold as a model for the development of Lake Monger Reserve.
9. Perth City Council, *Mayor's Report 1903–4*.
10. R. K. Clark, 'The City Beautiful: Promise and Reality', in *The Architect*, Vol. 10, No. 2 1969 and 'The Garden City Movement and Western Australia', in *The Architect*, Vol.10, No. 4 1969, discusses these two movements in relationship to Perth.
Perth City Council, *Mayor's Report 1909–1910*, under Parks and Reserves, gives a very comprehensive statement on proposals for development of the city, with recommendations for parks, the first children's playground, the removal of fences from parks and gardens, specifications for street trees, and a belt of parklands extending across the north of the city, in order that Perth may in future years worthily earn the name of the Garden City of Australia. An early reference to Garden City ideas occurs in the *Mayor's Report 1902–3*.
11. Clive Wainwright, 'Municipal Parks and Gardens', in John Harris (ed.), *The Garden: a Celebration of One Thousand Years of British Gardening*. New Perspectives Publishing Limited, in association with the Victoria and Albert Museum, London, 1979.
12. Howard Tanner and Jane Begg, *The Great Gardens of Australia*, The Macmillan Co. of Aust., South Melbourne, 1976.
13. *Inquirer*, 5 Dec. 1883, p. 5e.
14. M. J. Webb, 'Planning and Development in Metropolitan Perth to 1953', in *The City and Region of Perth, The Papers and Proceedings of the Tenth Congress of the Australian Planning Institute* (Perth; 1968), quoted from the speech of W. G. Brookman referred to in (1) above.
15. Clive Wainwright, 'Municipal Parks and Gardens'.
16. *West Australian*, 2 Aug. 1890, p. 5a. Jane E. Adams, 'Old St. George's Terrace', in *Early Days — Journal and Proceedings Western Australian Historical Society*, Vol. III, No. V.
17. *West Australian*, 9 April 1890, leader.
18. *Inquirer*, 2 Feb. 1876, p. 3a.
19. *Western Mail*, 8 Jan. 1948, p. 70, quoted in George Seddon, *Swan River Landscapes*, University of Western Australia Press, Nedlands, 1970.
20. Brent Elliott, 'Victorian garden design', in J. Harris (ed.), *The Garden: A Celebration of One Thousand Years of British Gardening*.
21. Perth City Council, *Mayor's Report 1897–1919*, including City Gardener's Reports.
22. J. Ednie Brown, *Annual Progress Report of Woods and Forests Department*, Perth, 1897–98. Perth City Council, *Mayor's Report 1909–10. Report of the Metropolitan Town Planning Commission Perth: Western Australia*, Perth, 1931.
23. V. Hughes, *History of Kings Park*. C. T. Stannage, *The People of Perth*.
24. C. T. Stannage, *The People of Perth*, 'Perth is a garden city', p. 4.